COASTAL ECOLOGY

Coastal

Ecology
Bodega Head

MICHAEL G. BARBOUR

ROBERT B. CRAIG

FRANK R. DRYSDALE

MICHAEL T. GHISELIN

Drawings by Mary Breckon

Photographs by Ted Barnes

551269

UNIVERSITY OF CALIFORNIA PRESS · BERKLEY, LOS ANGELES, LONDON

University of California Press
Berkeley and Los Angeles, California
University of California Press, Ltd.
London, England

Contents

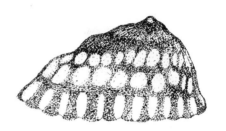

Figures and Tables

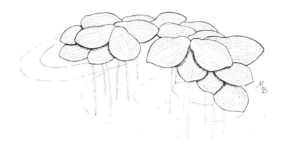

TABLES

Preface

This book can be read on several levels. We planned it that way because Bodega Head and places like it should interest a wide spectrum of people: professional biologists seeking data they can use in their research, amateur naturalists who want to see the "forest" *and* some of the trees, students in need of information to supplement their course work and field work, those people who simply want to appreciate the place where they live or vacation, and anybody who happens to enjoy reading books on natural history.

For the most part we deal very specifically with Bodega Head in northern California, its history, geology, climate, organisms, and what uses we make of it. By focusing on this concrete example, we hope to illustrate how some general principles of ecology and ecological research apply to a real ecosystem, and not just a textbook abstraction. When science has become so general in tone as to be pure abstraction, the time has come to bring it down to earth again.

We begin by explaining our point of view. Ecology is a science, and every science has a variety of methods and outlooks. There are all sorts of ways to "do" ecology, and our's is only one of them: "habitat ecology." We emphasize in this book the kinds of plants and animals that live in a given habitat, such as a marsh or a mudflat, and the environmental factors which affect them. We consider in turn each of the major habitats and a sample of the organisms found there, providing facts, theories, and speculations that should lead to a scientific interpretation of the underlying processes. We cover grassland, the rocky shore, beaches and dunes, the harbor and salt marsh, seeps and fresh-water marshes. Finally we discuss man's past and present activities at Bodega Head: primitive hunter and gatherer, farmer, rancher, fisherman, and advocate or foe of nuclear power plants. We have not attempted to tell the world how it should deal with Bodega Head, but what we say may help others to make sensible decisions about its management in the future.

Bodega Head is changing, and it will continue to change. Waves are eroding the cliffs, wind is reshaping the dunes and filling the harbor, plant communities change their composition—and animal life changes with them. Man's influence will also change in coming years, in ways that are hard to predict. We provide here a sketch of what we found at only one moment in time, hoping that our record will be clear enough for those who come after us.

We would like to express our gratitude to those people and institutions who have made our work possible. Warren Cothran identified insects; Archie Waterbury helped with mammals; Rose Gaffney provided historical information. All but two of the line drawings were beautifully and patiently executed by Mary Breckon. Sean Barry drew the shrew and microtus (Figs. 2.30, 2.34). Because of the book's schedule, both often had to make do with difficult material. The photographs were taken by Ted Barnes. Joel Hedgpeth contributed a great deal of technical advice on the manuscript. We owe a great deal to the Bodega Marine Laboratory and the persons associated with it for providing services, assistance, and a delightful place to work. We especially thank its Director, Cadet Hand, who helped us over a period of several years in ways too many to enumerate. Malcolm Erskian, the "Night Director," helped collect data and Eric Davidson has collected and identified many new plants during the past two years. Our personal store of knowledge has been greatly increased by the activities of students, teachers, and researchers who use the Laboratory. We are particularly grateful to those who permitted us to use materials from their doctoral research: David Cobb, Kenneth van der Laan, Eric Davidson, and Thomas Wolcott. We regret that limitations of space prevent our thanking individually many other people whose contributions are nonetheless deeply appreciated.

1
Introduction

Habitat Ecology

By definition, the science of ecology is concerned with the relationships between organisms and their environments. The discipline admits a diversity of outlooks and points of view, each of which contributes in its own way to our general picture of the living world. The present study deals with what is called "habitat ecology." By the "habitat" of an organism, we mean what is metaphorically spoken of as its "address," or where it lives, and the object of such a work as this is to describe and explain how life goes on in such a location. To this end, we consider the various environmental influences in a given situation, and see how the inhabitants cope with them.

We shall deal with a single, concrete example: Bodega Head in northern California. As in any scientific study, however, this single coastal habitat does represent something more general. An extensive part of the North American coastline

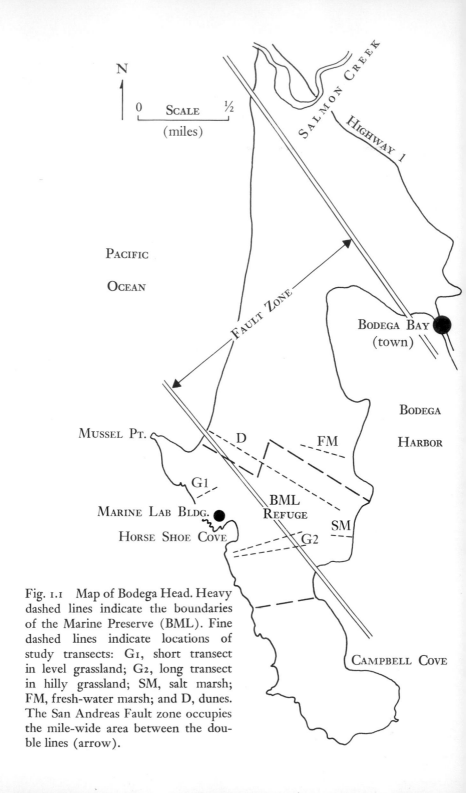

N

0   SCALE   ½
(miles)

PACIFIC

OCEAN

SALMON CREEK

HIGHWAY 1

FAULT ZONE

BODEGA BAY
(town)

BODEGA

HARBOR

MUSSEL PT.

D   FM

G1

MARINE LAB BLDG.

BML
REFUGE

SM

HORSE SHOE COVE

G2

CAMPBELL COVE

Fig. 1.1   Map of Bodega Head. Heavy
dashed lines indicate the boundaries
of the Marine Preserve (BML). Fine
dashed lines indicate locations of
study transects: G1, short transect
in level grassland; G2, long transect
in hilly grassland; SM, salt marsh;
FM, fresh-water marsh; and D, dunes.
The San Andreas Fault zone occupies
the mile-wide area between the double lines (arrow).

has a climate and an assemblage of plants and animals much like that studied in our very limited area. And the meeting of land and water over much of the globe produces comparable effects. Indeed, the zone of contact between any two habitats ("ecotones" as they are called) has much in common with any other such meeting place. This is not to say that the particular examples are uninteresting as such. When we see an actual natural situation, we appreciate the uniqueness and variability of the real world. We see how the abstract principles are applied, and at the same time realize their limitations.

The habitat approach helps to provide a secure foundation for other kinds of ecology, and we shall attempt to fit some of these into our discussion. When we have considered how various environmental influences, such as salt spray and wind, are distributed in space and time, we then have a basis for analyzing how each kind of organism is adapted to them ("autecology" or the study of how an organism relates to its environment). And when we view all the organisms together as a functioning whole, we gain some conception of an ecological community ("synecology"). It becomes possible to treat each kind of animal or plant as part of a "natural economy" in which it has a "profession" or a "niche" in one sense of that term (see MacArthur 1968). Finally, when we add to the communities of organisms the non-living parts of their environments such as sunlight and air, we can build up to a yet more encompassing level: that of the "ecosystem."

Bodega Head

Bodega Head is a small coastal peninsula about 65 miles north of San Francisco (Figs. 1.1, 1.2). Bounded by State Highway One on the northeast, Salmon Creek on the north, the Pacific Ocean on the west and south, and Bodega Harbor on the east, its area of three square miles has a rather complex topography. On the southwest corner is a region of granitic rock, hilly and edged with steep cliffs along the shoreline. The term "Bodega Head" sometimes refers only to this part of the peninsula.

The granite itself is a quartz diorite that was formed about 100 million years ago, and which has a very peculiar distribution in California. It is common west of the San Andreas Fault, a zone of crustal movement running north and south through much of California. The same rock forms part of Point Reyes, Inverness Ridge, the Farallon Islands near San Francisco, and the Santa Lucia Mountains. To the east of the fault, the closest major area with similar granite is in the Tehachapi Mountains, about 300 miles south of Bodega Head. This east-west discontinuity could be explained as the consequence of land having slipped literally along the fault; this would suggest that Bodega Head once lay parallel with the Bakersfield area, but has since been shoved northward, along with an entire mass of land called the "Salian Block." But geologists are still uncertain that this is the best explanation for the discontinuity. At a recent conference a number of contributors pointed out that on the basis of soundly established geological evidence for movement, one can account for no more than a lateral slip of 160 miles at most over the last 25 million years. (This averages to about half an inch per year; Anderson (1972) shows however that the rate of movement during the last 200 years has averaged 2.5 inches per year.)

The San Andreas Fault begins in the Gulf of California and follows a sinuous path roughly toward the north-northwest through San Francisco and Salmon Creek. It finally turns west out to sea off Cape Mendocino in northern California. Along this 700-mile stretch the fault zone may be as narrow as ten feet, or as wide as several miles. At Bodega Head it is about a mile across, and corresponds to the sand dune area between the grassland and Salmon Creek (Fig. 1.2). Land movement along faults is associated with earthquakes, but not all parts of the San Andreas Fault are equally active. Earthquakes

Fig. 1.2   Aerial view of Bodega Head

are most frequent in a southern California region from about Santa Barbara south, and in a section of central California from Parkfield north to San Francisco. The fault is not particularly active at Bodega Head. Damage to existing structures there during the 1906 earthquake—which devastated San Francisco—was minor, although the ground moved laterally five to ten feet.

The grassland reaches a higher elevation (266 feet) than does the dunes area. A pronounced foredune, less than 40 feet high, forms an inland boundary to the beach. Even the hinddune, which parallels the foredune much further inland, is less than 150 feet high. The northeastern portion of the peninsula is built of marine deposits which are no more than several million years old. Several gulleys cut through its low hills and run roughly from east to west.

The soils of Bodega Head are predominantly sandy, ranging from pure sand in the dunes to a loamy sand in the grassland, where the soil is darkened by considerable organic matter from decaying roots and herbage. About 45 acres along the harbor side are occupied by a low, fresh-water marsh. Many fresh-water seeps open to the surface along the steep hillsides and cliffs.

Many Indian artifacts have been found on the Head indicating that it has been occupied by man for some time. The Spanish explorer Juan Francisco de la Bodega y Quadra discovered Bodega Bay on 3 October 1775, and produced a chart of the bay and harbor. The first Europeans to settle the immediate area were Russians. In 1809 Ivan Kuskov built a settlement at the southeast side of Bodega Head, near the inlet at a place now known as Campbell Cove. The site, he noted, was wind-blown and lacked trees, but it did have a fresh-water spring.

The Russians built a larger settlement at Fort Ross in 1811. The area between Bodega Head and Fort Ross was devoid of timber, but it proved adequate for grazing and vegetable farming. Potatoes soon became the favorite crop, and cattle, sheep, horses and pigs, purchased from the Spanish to the south, grazed the area heavily. By 1830, Fort Ross had become established as a shipyard, while the settlement at the Head was an import-export center for food, bricks and general supplies. Ten thousand bricks were shipped from Bodega in 1830, and the harbor became known to Americans as a good place to anchor and take on water.

Within the next decade, however, the Russians had hunted the sea otter and seal nearly to extinction, and their California settlements came to be more of a liability than an asset. The local agriculture could not sustain the colony without imports, and it seemed impossible to negotiate with Mexico for more land. The Czar gave them permission to move to Sitka, and they arranged to sell all movable property to Captain John A. Sutter of Sacramento. Heavily in debt, Sutter never paid them, but the discouraged Russians left anyway in 1841. That year one of them compiled an excellent checklist of 214 plant species which grew along the coast between Bodega Head and Fort Ross. He coupled the list with a dried specimen of each species, and his collection is still preserved in the huge Leningrad Herbarium (Howell 1939).

Although Russia claimed never to have sold the land itself, Mexico "accepted" it, and divided it into large ranches. Rancho Bodega occupied some 35,000 acres along the coast from the Russian River to the Estero Americano River (the present boundary of Sonoma and Marin Counties). In 1844 Rancho Bodega was granted to an American, Stephen Smith, who soon built a sawmill in a redwood area to the northeast of the village of Salmon Creek.

The goldrush, statehood and the transcontinental railroad swelled the population, and Rancho Bodega was broken up; there came to be several holdings on the peninsula alone.

Among these, the Gaffney family owned over 400 acres in the center. Mrs. Rose Gaffney, who arrived on Bodega Head in 1913, still lives in Salmon Creek. She recalls that much of the present dune area was pasture in the early 1900s, but sand continually encroached from the west. The land was used primarily for potato farming and the grazing of dairy herds until the 1930s, when many marginally profitable herds were replaced by sheep. Mrs. Gaffney claims that sheep were grazed on her property for only two months in 1942, but that cattle and horses were regularly present. She remarks that since the property has become a preserve and grazing has been discontinued, the show of early summer flowers is much more spectacular than at any previous time.

The Pacific Gas and Electric Company purchased over 225 acres at the southern tip of the Head in 1959, as a site for a nuclear power plant. Their plan met with considerable opposition, and was ultimately abandoned (as we shall recount in Chapter 7), but not until a road had been built across much of the mudflat and a large hole had been excavated. As this book goes to press, negotiations are under way to take this land into the state park system.

In 1959, a Chancellor's Committee at the University of California's Berkeley campus recommended Bodega Head as a site for a marine laboratory. The University acquired 326 acres of land—most of the Gaffney property—in 1962, and established it as a preserve. A building was constructed, and the Bodega Marine Laboratory was opened in 1966. The Laboratory is a university-wide facility and although primarily established for the benefit of faculty and students from its nearby campuses, it provides services on a nation-wide, and even an international, scale. At present it is mainly used for teaching and research at the nearest branches of the University of California, Berkeley and Davis.

In 1962 most of the dune area along the west side of the

peninsula was purchased by the State of California and incorporated into Sonoma Coast State Beach. The remaining portions, along the harbor and Highway One, are privately owned.

The Climate and Physical Environment

Coastal habitats as a rule have a more moderate climate than do comparable regions inland. The reason has to do with the well-known physical properties of water: it holds a great deal of heat, and becomes warmer and colder much more slowly than does air. A large body of water will substantially influence the conditions on the adjacent land. During summer the slowly warming water cools the air above it, and even the gentlest breeze from the sea will depress the daily high temperatures on the land. In winter, the water loses its heat gradually, and the sea breezes then tend to increase the land temperatures.

The climate at Bodega Head can be characterized as cool and wet in winter, cool and dry but foggy in summer (see U.S. Dept. of Commerce 1968; Eber et al. 1968). In Table 1.1 the general climate of the Head is compared with that at other points along the West Coast. Toward the north it rains more, but solar radiation, air temperature and water temperature all decrease. Seasonal fluctuations in temperature reach a minimum near Eureka because of upwelling: a cold current rises to the surface offshore. Here the mean monthly air temperature changes only 10°F between the coldest and warmest months, and water temperature changes only 2°F.

The moderating effect of the ocean on nearby land is dramatized in Fig. 1.3, which compares Bodega Head with Sacramento, some 90 miles inland. Sacramento not only has a higher mean annual temperature, but the temperatures fluctuate more. Similarly, the mean minimum temperatures are

Location, °N Lat.	Sunshine (% of possible)	Rainfall (inches)	Air Temperature			Water Temperature		
			Mean of coldest month	Mean of warmest month	D	Mean of coldest month	Mean of warmest month	D
San Diego, 32°40'	70	10	55	72	17	57	65	8
Morro Bay, 35°20'		17	52	65	13	55	60	5
Bodega Bay, 38°20' *	61	31	51	63	12	53	56	3
Eureka, 41°00'	49	38	47	57	10	51	56	2
Newport, 44°35'		66	39	58	19	49	59	10
Juneau, 58°15'	31	55	25	55	30	42	56	14

* Weather data come from Fort Ross, about 16 miles north of Bodega Bay, except for sunshine which comes from San Francisco and rainfall which comes from rain gauges at the Marine Laboratory.

Table 1.1 Climate along the Pacific coast of North America. (D = difference in temperature between means of warmest and coldest months.)

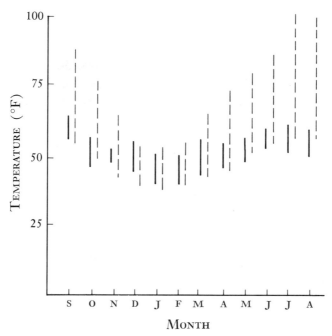

Fig. 1.3   Mean monthly maximum and minimum air temperatures on Bodega Head (solid lines) and at Sacramento (dotted lines) for 1969.

lower in winter, the mean maxima higher in summer. The range of temperature fluctuation for the entire year is three times as great in Sacramento as it is on the Head.

The amount of sunlight entering an area is important, partly because it helps to determine the temperature, and also because plants with photosynthetic pigments transform much of its energy into food. The amount of this solar radiation can be measured with a small device called a pyranometer (see Appendix B), and expressed as calories of energy per unit area per unit of time. The seasonal change in solar radiation at Bodega Head is shown in Fig. 1.4. Many factors affect the amount of solar radiation which reaches the plants on the ground or in the water. Daylength is one of the most obvious:

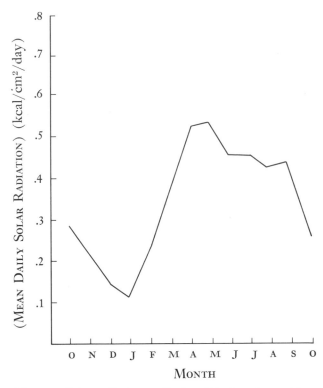

Fig. 1.4   Solar radiation on Bodega Head, 1969–70. Measurements were made with a pyranometer.

in winter the days are shorter and the sun is at a lower angle than in summer, and consequently, less energy reaches the ground. Winter storms bring extensive cloud cover that further reduce incoming radiation, and, especially in summer, fog has the same effect. Peak solar radiation at the Head was about 0.50 kilocalories (kcal) per square centimeter (cm) per day in June; the minimum was 0.15 kcal/cm²/day in January. The average for the year was 0.35 kcal/cm²/day, which is about twice the yearly average for such continental areas as central Michigan.

Rainfall at the Head is seasonal, over 90% of it coming

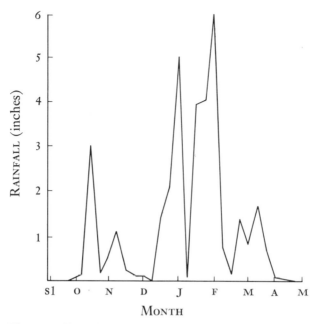

Fig. 1.5 Weekly rainfall on Bodgea Head, 1969–70.
Over 90% of all rain falls during the period October–
April.

in the period from October to April. This does not mean that
winter months are continuously wet: wet and dry weeks succeed each other seemingly without pattern (Fig. 1.5). Following the first rains in October and early November of 1969, for
example, there was about a four-week drought, and in January and February of 1970 there were further one-week
droughts. As will be shown in later chapters, these intervals
of aridity affect the germination survival of some plants.
(Rainfall was collected and measured in rain gauges placed
on the Marine Laboratory roof—see Appendix B.)

Another form of moisture is fog. Even though the great
bulk of a plant's water needs are met by soil water absorbed
through the roots, there is evidence that some plants can use
atmospheric moisture which they take up through their leaves.

At the least, the plant suffers less water stress when high humidity, fog or dew reduces the amount of water passed to the air. Ponderosa pine is one plant that may utilize fog or dew. A common forest tree in western mountains, it sometimes grows far down slopes, where soil moisture can be low for several months. The Berkeley forest ecologists Stone and Fowells (1955) conducted an experiment with this species, in which one-year-old seedlings, growing in pots of soil, were exposed to fine mists every night. A control group was not exposed to mist. Both groups of pots were surrounded with waterproof material, so that spray could fall on the foliage but not on the soil. When the soil moisture was depleted, the seedlings died, but the group exposed to mist lived, on the average, one month longer than the control group. The results indicate that the plants could use atmospheric moisture to complement that in the soil.

Another way in which plants may utilize fog is by condensing the mist into large drops which then trickle down the stem or drip from the branches onto the soil below, where they can be taken up by the roots. G. T. Oberlander (1956) placed catch basins beneath trees and found that on the San Francisco peninsula, depending on the type of exposure, the amount of water added to the soil in this manner could be more than twice the amount contributed by rain.

Fog is mainly a summer phenomenon at Bodega Head. The results of visual observations, averaged from a two-year period, are shown in Fig. 1.6. Fewer than 10% of winter days have fog, while up to 80% of summer days are foggy. Often the fog lies along the ocean-facing grassland and dunes until around 10 A.M., clears away, then returns in the late afternoon. If one cannot be present every day to observe the fog conditions, one may wish to use a fog gauge such as that described by Vogelmann and his collaborators at the University of Vermont (Vogelmann et al. 1968). A cylinder of fine mesh

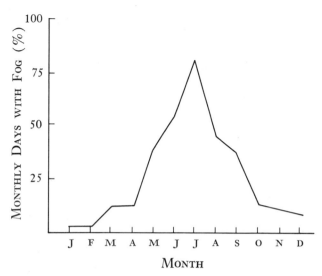

Fig. 1.6   Incidence of fog on Bodega Head. Figures are averages of 1968 and 1969.

wire is fitted to the mouth of a rain gauge so that it extends about six inches above the mouth. Passing fog will condense on the wire and accumulate in the gauge. A rain gauge without the wire is placed nearby, and the difference in water level between the two indicates the portion contributed by fog.

We shall see in later chapters that fog affects the distribution of plants at Bodega Head, but its effect may be due to salt rather than moisture. In a classic study, Stephen Boyce (1954) showed that oceanic fog condenses around tiny droplets of sea water. Turbulence, such as that due to breaking waves or whitecaps, forms bubbles of foam, which burst and shoot the droplets of sea water into the air. The height to which the droplets are shot depends on the diameter of the bubble. They seem to be shot no more than 20 centimeters, but this is enough for the droplets to be picked up by slow winds near the surface of the water and mixed into higher air. If the air is saturated with moisture, pure water will condense around the droplets. Fog trapped on the Head in fog gauges is brackish to the taste; it can corrode cars, buildings and plants alike.

Winds on the Head blow predominantly from the north-west. Wind speed has been measured with a cup anemometer mounted about 30 feet off the ground (see Appendix B) and recorded continuously on chart paper. This instrument indicates that wind speed averages 8 to 10 mph. Storm peaks commonly reach 40 mph, and stormy January is the windiest month. The winds have caused erosion in the dunes area, and sand thus eroded has ultimately made its way to the harbor, making it necessary to dredge the ship channel. In an effort to reduce erosion, beach-grass has been planted in the dunes a number of times since 1900. These plants, introduced originally from Europe, seem particularly well adapted to rapid growth in sandy areas. Their large fibrous root systems help to anchor the sand, and their narrows leaves resist the sand blast, which is powerful enough that it removes paint from signs.

Wind may carry a substantial amount of salt onto the land. On calm days salt is not carried far inland, but storms may transport it considerable distances. Salt spray droplets have been detected in the air hundreds of miles from the nearest coast, and one would suspect that they could contribute substantial amounts of nutrients to the soil. One of the common elements in salt spray is chlorine. More than half a century ago W. K. Knox (1915) measured the amount of chlorine carried down in rain and snow during a nine-month period at Mt. Vernon, Iowa. At this point, over 800 miles from the nearest seacoast, he found that nearly 37 pounds of chlorine were deposited per acre. Chlorine is so ubiquitous in the air, and so difficult to exclude, that until recently it was not suspected to be a required element for plant growth. Several University of California plant scientists (Ulrich and Ohki

Period	1968 Range	1968 Mean	1969 Range	1969 Mean	1970 Range	1970 Mean	1971 Range	1971 Mean
Jan.	9.2–11.4	10.2	10.6–11.6	11.0	10.8–13.1	12.3	9.6–11.1	10.3
Feb.	10.4–13.5	12.1	10.5–11.6	11.2	12.3–13.2	12.6	8.2–10.2	9.5
March	10.3–13.5	12.1	9.7–14.6	11.4	9.6–12.8	11.4	7.7–10.7	8.9
April	8.3–12.7	9.9	9.8–13.4	11.2	7.9–10.1	8.8	7.8– 9.9	8.8
May	9.7–12.5	10.6	9.3–13.5	11.2	7.8–10.6	9.2	8.3–10.4	9.1
June	9.0–14.2	10.8	10.6–14.4	12.5	9.0–11.8	10.5	8.1– 9.9	8.9
July	10.0–13.3	11.4	10.1–15.4	12.4	9.3–14.5	10.7	8.8–11.2	9.6
Aug.	10.5–14.5	12.6	10.7–14.1	11.9	9.8–11.5	10.8	9.3–14.1	10.7
Sept.	10.8–15.0	12.6	12.0–15.2	13.4	9.9–12.8	11.0	10.6–14.1	12.3
Oct.	11.8–13.4	12.6	10.7–13.9	13.2	9.7–12.8	11.6	9.3–12.0	10.8
Nov.	11.1–13.3	12.3	12.1–13.8	13.2	12.5–13.5	13.1	9.8–11.0	10.3
Dec.	10.3–11.4	11.1	12.5–14.1	13.3	10.8–12.9	11.9	8.8–10.5	9.5
Year	8.3–15.0	11.5	9.3–15.4	12.2	7.8–13.5	11.2	7.7–14.1	9.9

Table 1.2   Surface seawater temperatures at Horseshoe Cove.

1956) found that they could demonstrate the need for chlorine in tomatoes and sugar beets only when they took the unusual precautions of growing the plants in controlled greenhouse environments and filtering the incoming air with activated carbon to remove all chlorine. Plants grown under such conditions grew poorly, and developed disease symptoms which could then be corrected by adding a trace of chlorine to the watering solution.

Sea water itself is a remarkably stable environment, at least in comparison to air. Temperatures in the open water off the Head are low (it is too cold for swimming without a wetsuit), and they remain impressively constant throughout the year. Temperature has been monitored at the Bodega Marine Laboratory since 22 May 1967; these records have been taken at the sea water intake, which occupied an exposed position, on week-day mornings. Data for the four-year period of 1968 through 1971 have been tabulated in Table 1.2, giving the maximum, minimum, and mean for each month in degrees Celsius (Centigrade). We have put up with the usual practice of giving air temperatures mainly in Fahrenheit and sea water temperatures in Celsius, except where we are trying to compare the two media. (To convert Celsius to Fahrenheit multiply by 9/5 and add 32.) During these entire four years, the lowest temperature recorded was 7.8°C, the highest 15.4°C. This gives us extremes 7.6°C apart, but it should be noted that the greatest range in a single calendar year (recorded in 1968) was only 6.7°C.

The greater thermal stability of water relative to air is again apparent when we consider the time at which the minima and maxima occur. Sea temperatures bottom out some time in April or May, not January or February as on land (cf. Fig. 1.3). They then begin to rise, and may not reach their maximum until well into the fall; and the subsequent decline occurs gradually. The maximum solar radiation level actually coincides with minimal water temperatures. On a year-to-year basis, we observe modest variations, much as we do on land, but unusual "weather" in the water itself deviates less from the norm than on land.

The cold, stable temperatures, as we have mentioned, are fairly representative of what occurs all along the coast from Point Conception northward. Such conditions, by no means unusual, do contrast strongly with those on the East Coast of the United States; at Cape Cod, for example, the water is warmer in summer and colder in winter than at Bodega Head. Although the low temperature would seem to exclude many species native to warmer waters, it actually has little effect on the suitability of the water as a habitat. G. M. Smith (1945) points out that the diversity of marine algae on the West Coast is far greater than it is on the East Coast: the number of species around the Monterey Peninsula alone approximates that for the entire Atlantic Coast from Hudson Bay to Virginia. Here seaweeds grow luxuriously, and the productivity of benthic algae is further seen in the remarkable abundance and diversity of animals which eat them, such as limpets and chitons. The constancy of the temperature may help to explain the variety of organisms, but the absolute quantity of life is probably best attributed to the fertility of the water. The sea water forms a nutritious organic soup, containing numerous small plants and animals; and one of the most conspicuous elements in the fauna of rocky shores are the "filter" or "suspension" feeders which support themselves by straining this food out of the water. The growing season for marine organisms at Bodega Head is obvious to anyone who looks for it. Since the increase in activity begins quite early in the year, the amount of solar radiation, rather than the temperature, would seem to be responsible for the seasonality. Upwelling due to wind may also be important, because this would bring nutrients up from deeper water.

The stability just described applies only to the more open waters. Isolated, or partly isolated, parts of the sea may become much warmer or colder, affecting the life of the inhabitants. Tide pools, for example, may become quite warm to the touch. Similarly, the water in Bodega Harbor is only in partial communication with the open ocean. At Bodega Head the limpet *Acmaea limatula* is quite abundant on rocks inside the harbor, but is only rarely encountered on the wave-swept outer coast. This species is said to become more common on the outer coast toward the south (Test 1946). The variation in sea water temperatures is complicated by the tidal patterns. We shall defer discussing such matters until later chapters.

Community Structure

The coastal habitat consists of six major units: grassland and ocean-facing cliffs; rocky intertidal zone; beach and dunes; salt-water marsh and mudflat; fresh-water marsh and seasonally wet areas; and, finally, disturbed areas near paths and buildings. Each of these habitats occupies a more or less discrete area, and each is populated by a distinctive assemblage of organisms. These inhabitants may be thought of as forming a "community" within each habitat. These communities may be characterized by drawing up species lists showing clear differences in which organisms live where. And it is evident too that many important biological interactions go on exclusively between members of the same community: thus animals which can live only in the burrows of mudflat animals are themselves restricted to the mudflat. Yet we could arrange our materials in a somewhat different fashion. We could subdivide the beach and dunes, or unite the latter with the grassland. Lupines and their associates may be found on both grassland and dunes. Furthermore, some animals regularly move from one habitat to another: marsh hawks and deer, for example. Food, water and nutrients pass from one part of the Bodega Head ecosystem to another, as well as into and out of it—which is only to be expected, since every such unit forms an integral part of a larger whole, the "biosphere" or world ecosystem. Finally, clear as the boundaries may seem, they are far from being altogether distinct: even the break between land and sea shows a gradient in the intertidal zone.

Such considerations raise the age-old philosophical issues that trouble every attempt at scientific classification. Are we, in singling out a community or other assemblage of organisms for study, dealing with a "mere abstraction?" Or, to the contrary, have we focused upon something which has been discovered, and which exists apart from our conceptions? The two views are not mutually exclusive, as we shall see. In ecology two extreme positions have been maintained on the reality of groups, neither of which seems adequate, but both of which contain a small measure, at least, of truth. On the one hand we have the "individualistic" conception championed by Gleason

(1926, 1939), who denied the "reality" of communities, and sought to explain the distribution and abundance of organisms as the result of each individual responding to the environmental forces that act upon it. At the other extreme, we have the "super-organism" concept, favored by many ecologists (especially Clements 1916), which treats communities as if they were every bit as concrete, particular things as you and we. According to this view, a community would have a life-cycle (succession) and the components of the community would stand in relation to each other as do the organs of the body.

What truth there is in the "individualistic" conception is easily seen in the present work. Indeed, one value of a habitat approach is that it helps us get an appreciation of the importance of individual organisms. For Bodega Head, the point will be made by observations on *spatial zonation*. Environmental influences, such as temperature, salinity and the availability of water, show a clear spatial gradient. In moving inland from the coastline one may see that the different kinds of organisms gradually replace one another, so that within the grassland community the composition at each end of the gradient is quite different. Species likewise replace each other in radiating circles about fresh-water marshes. Nowhere, perhaps, is zonation more conspicuous than on the intertidal rocks, where bands of differing kinds of animals and plants are seen at a glance. Nor should we fail to note the gradual progression of species in time, in the sense that there are different seasons of growth and of other activities. Yet none of this implies that all is arbitrary. It only means that within each community different species are adapted to sections of gradients within the habitat: the component individuals are able to withstand their competitors and handle the environmental circumstances in a limited part of the range. Such a subdivision of broader niches will be illustrated in Chapter 3, when we show how the broad "profession" of the limpets is practiced in a somewhat different manner by the species that live at different levels in the intertidal zone. But limpets remain characteristically intertidal animals all over the world. Furthermore, many of the associations we observe in nature have been produced by long periods of evolution, and the organisms have come to depend upon one another; hence, their associations are influenced as much by their usual neighbors as they are by physical and chemical influences. We see such interrelationships quite nicely in mussel beds. A host of small animals make their home between the mussels, and their characteristic limitation to this position suggests that the mussels themselves determine the distribution of their associates.

In agreement with the "super-organism" point of view, we could treat each community as a system of interdependent units with a definite structure. The occupants of each niche could be thought to have a function subserving the needs of a larger whole. Thus, each species would occupy a more or less definite position in a *food chain*, that is to say, in a sequence in which energy and materials get passed from one kind of organism to another. Such a chain is generally thought to involve three basic kinds of organisms: *producers, consumers* and *decomposers*. Every community must of necessity have a source of food: the organisms which provide it are the producers. Usually the producers are plants which generate energy-rich compounds from sunlight, but the mussels in a mussel bed are also looked upon as producers, since they eat plants that come to them from the water rather than from the community itself. Herbivores and carnivores would be the "consumers," ultimately deriving their food from plants. The "decomposers" are organisms which eat dead plant and animal materials—bacteria, fungi, scavenging insects, vultures, and the like. They have the important effect of breaking down the materials locked up in the food they eat, and making it available once more to the producers. Of particular importance are nitrates and phosphates, known to be in short supply on both land and sea; but other elements, such as silicon in the sea and various trace elements on land, are also important to some groups of organisms. Although some nutrients are constantly being added to soil and to marine water, through decay of rocks, maximum growth of plants and animals depends upon a continual recycling of nutrients. That this is so is attested to by the fact that over the majority of the earth's surface, nutrients are decomposed largely after having been carried down into the great depths of the ocean, so that the surface water (where photosynthesis occurs) is far less fertile than it could be. Where, as on the coasts of California, West Africa and Peru, water rises from great depths, the productivity is unusually high, but such conditions are exceptional.

We should not, however, view the producers as if they really were, like our own digestive apparatus, making food "for" the other species. Neither is a food chain equivalent to a circulatory system, nor a vulture to a kidney. Communities are not cooperative associations like organisms, but competitive ones, more like laissez-faire capitalist economies. The super-organism concept in its traditional form is a remnant of the old-fashioned "teleological" conception of nature, which presupposed that this is the best of all possible worlds, even viewing horses as something which existed in order that we could have transportation. Popular writers, theologians and even the authors of certain textbooks still maintain it, exaggerating the idea of a "balance of nature" and treating natural situations as if they were ipso facto "good" or "optimal." Such notions are as misleading as they are unscientific.

It seems best, then, to view a community as concrete, but more as the counterpart of a society rather than an organism. A loosely-integrated whole, with nonetheless interdependent parts, its component organisms have adaptations which they have inherited from their ancestors; they live where they can, both because of their neighbors and in spite of them. Just how harmonious, integrated or balanced an eco-

system may be deserves more investigation. But in finding out
what actually determines the structure of a community, we
should look, not to analogies, but to the organisms themselves
and to their habitats.

The Scientific Method in Habitat Ecology

One reason for writing this book has been our belief that
every science should rest upon a solid basis of fact. To this
end we include a considerable body of descriptive material,
much of it new. The major kinds of plants and animals are
inventoried, and data are presented on their abundance and
distribution. We have compiled a nearly complete checklist
of all vascular plants and vertebrates known from the Head
(Appendix A). These lists summarize common and scientific
names or organisms, their habitats, abundance, seasonality, and
miscellaneous observations. Some of the more common or interesting forms are illustrated here, but a complete pictorial
guide would have been prohibitively expensive. The interested
reader may wish to consult the general references listed at the
end of the book for additional checklists and aids to classification. The main goal of the descriptive material, however, is
more than a catalog or guide. Rather the aim is to show how
and where a representative sample of organisms do, in fact,
live.

Given such facts, it should be possible to analyze and to
explain, in terms of theory, what actually determines the
structure of some real communities. In so doing we may hope
to avoid the excesses of "pure" theoreticians who construct
"models" having no counterpart in objective reality. This is
not to say, however, that theory has not entered into what we
have done. Far from it. Ecologists have a pretty good idea of
what is likely to be going on at Bodega Head and elsewhere.
The data are not collected at random, but rather in a manner
so as to see whether these ideas apply to the situations being

studied. Thus, it makes a great deal of sense, in studying zonation, to sample along a gradient at more or less right angles to the shore, because this is the direction in which the gradient runs. Likewise, we have every reason to measure those influences, such as temperature and salinity, which do in fact vary from one place to another and which do affect the lives of organisms. We would be wasting our time measuring the percentage of the inert gas Argon in the atmosphere.

A lot of theory inevitably goes into what on the face of it is a "purely descriptive" study. But one should not get the impression that our conclusions are derived from theoretical assumptions. The sampling program is set up to test *hypotheses*. That is to say, we ask "Is it true that zonation is related to temperature?" We thus formulate two hypotheses on the basis of this question: 1) zonation is related to temperature, 2) zonation is not related to temperature. It is logically impossible for *both* of these hypotheses to be true, and the right kind of observations should let us decide in favor of one or the other. The zones will, or will not, coincide with the gradient in temperature. We need only measure and observe.

Such correlations, as is well known, have certain drawbacks, especially in establishing causality. Consider what happens when we cast the hypothesis in different terms: "Animals are restricted to their positions in zones because they can survive only at the temperatures where they live." Showing that they can live elsewhere refutes this notion, but showing that they are restricted to a particular temperature does not confirm it, for something else might be limiting their distribution. A good example may be taken from a study by Connell (1961), who studied two kinds of barnacles on the coast of Scotland. One of these, *Chthamalus*, lives higher up where, indeed, it is warmer, and one might infer that a higher temperature was necessary for its existence. It turns out that if the other barnacle, *Balanus*, is removed, the *Chthamalus* does quite well further down. When both are present the *Chtha-*

malus gets forced out by "competition" from the *Balanus*, which grows faster and crowds the other out. Such experiments illustrate a logical principle which, however basic, always bears repeating: we can refute hypotheses, but not "prove" them for sure.

The difficulties of comparison and experiment become even more apparent when we try to explain our data in yet more fundamental terms. It is becoming increasingly evident that competition, comparable to that which Connell has documented for barnacles, plays an important role in determining what lives where. And it seems to have contributed, during evolutionary history, toward molding the diversity which we now see in contemporary biotas. According to the traditional notions of Gause (which very likely are not strictly correct), two species cannot occupy "the same niche." Should two species with identical niches come to live in the same place, two results are possible. The first possibility is "competitive exclusion": one species becomes extinct. Alternatively, they may evolve differences in their way of life, and, as it is said, subdivide the niche. This is all very hypothetical, but the fact seems clear that closely-related species do diverge when they live in the same place, tending, for instance, to adopt slightly different feeding habits.

In following chapters we shall repeatedly see how certain niches have been subdivided along gradients, so that different species, often fairly closely related ones, replace each other in zones. It would be very satisfying if we could tell, from these facts, what has brought a given species to occupy a particular position. Alas, we can as yet derive very little insight from such data. We can tell that one species excludes another in a given circumstance, but we cannot yet say why some niches have become divided in the course of time, while others have been monopolized. We shall have to content ourselves, therefore, with showing "how" rather than "why" such patterns are apparent.

2

The Grassland
Ecosystem

Some Questions

The word "grassland" might bring to mind the region which greeted westward-bound pioneers when they reached the midlands of America: a vast, gently rolling expanse covered with tall grasses, with an occasional cluster of trees in the hollows of hills. The soil beneath, black with organic matter from decayed fibrous grass roots, proved very fertile, ideal for growing corn and wheat. Today, almost all of the American grasslands have been converted to farm or pasture, and plant ecologists must rely on the notes of early explorers, boundary surveyors, and settlers, and on their imaginations, to reconstruct what that vegetation type looked like.

The Bodega Head grassland (Fig. 2.1) differs in a number of ways from that stereotype. First, it is very restricted in

Fig. 2.1 Grassland aspect, showing rocky outcrops. The Bodega Marine Laboratory is in the background.

extent, occupying only the southern portion of the peninsula, an area of about one square mile. Nor is it as homogeneous as the plains; it covers a complex topography of precipitous cliffs and gulleys facing the ocean, rolling hills, rock outcrops on exposed hilltops, sheltered flats, and wind-swept points. Finally, the prairies were dominated by perennial species native to North America; most of the plants in the Bodega Head grassland are annual, and one-third of them are forms that have been introduced from Europe during the past 200 years.

The grassland ecosystem has been singled out for study in our research and for answering a wide variety of questions. We have sought descriptive materials. What species of plants and animals are found there, and how are they distributed? Are there any seasonal differences in their behavior? What are the physical factors of this environment, and how do they change in time and space? We have also asked questions demanding analysis and experiment. Why are particular species limited to narrow habitats within the grassland, and how do environmental factors influence their behavior? What are the interrelationships between species? How has the grassland been modified by man's activity? Finally we have considered what is the future composition of the grassland likely to be, given freedom from disturbance?

A Quick Look at Zonation, and at How to Document It

Considering the topographic diversity of the grassland, it is not surprising that many species are restricted to only a portion of the area. Among the more abundant plants, sea pink, *Armeria maritima*, is found only along the lip of ocean-facing bluffs; polypody fern, *Polypodium scouleri*, is restricted to rocky hilltops; and lupine, *Lupinus arboreus*, and fiddleneck, *Amsinckia menziesii*, are most common on the lee side of hills and in protected flats. The same applies to animals: the vagrant shrew, *Sorex vagrans*, prefers dense grass cover and is rare on rocky hilltops or beneath lupine shrubs

Fig. 2.2   Sea pink

where grass is thin; however, the marsh moth caterpillar, *Estigmene acraea*, subsists on lupine buds and so is closely associated with lupine wherever it occurs. These species, then, occupy different zones or regions of the grassland habitat.

Many grassland species also occupy different zones of time. In early spring the grassland is a lush, green color, dominated by many small herbs such as miners' lettuce, *Monita perfoliata*. In May, it becomes carpeted with yellow-petaled flowers, principally goldfields, *Lasthenia chrysostoma*. By midsummer, many of the grasses such as the annual Italian ryegrass, *Lolium multiflorum*, flower in abundance and begin to turn light brown. In fall, the grassland is a monotonous, dull gray-brown color from all the annuals which have died in the summer drought, except for interrupting green plants of bull thistle, *Cirsium vulgare*. At this time of year the vegetation pleases the eye far less than it does in the spring. Seasonal variations are also common for animals, as we shall see later in this chapter.

The fact that this zonation of species in time and space correlates with zonation or gradients in the microenvironment can readily be apprehended. One has only to stand on the lip of the bluffs, then walk back over a hilltop to the protected flats, to feel large differences in temperature; or one need but

Fig. 2.3   Polypody fern

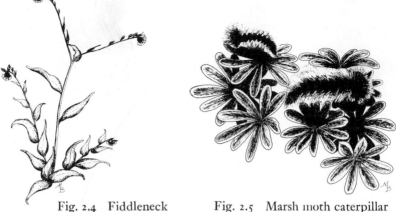

Fig. 2.4 Fiddleneck   Fig. 2.5 Marsh moth caterpillar
on lupine leaves

visit the Head at different times of year to observe seasonal differences in rainfall, temperature, or incidence of fog. But as we said earlier, broad correlations like these don't really contribute much to a fundamental understanding of how an ecosystem works. Have any other environmental factors been left out? What are the exact distributions of the species, and how fast do they appear or disappear with time and distance? How does each species affect the others, by providing food, by influencing the microenvironment, or by predation? The answers to these questions require measurements and experiment.

One can obtain detailed field measurements of a gradient in the environment by sampling the relevant influences at regular intervals along a line that parallels the gradient. Three such sampling lines (transects) were established in the grassland at Bodega Head (Fig. 1.1). A short transect was located on level grassland just to the north of Horseshoe Cove, extending 50 m back from the lip. The environment was sampled every 10 m along this line. Two long transects were located on hilly terrain just south of Horseshoe Cove, each one extending 780 m back from the lip. These two transects were roughly parallel to each other: one crossed the very top of a hill, the other crossed its shoulder. The environment was sampled here principally 15 m in from the lip (called the lip position) and 310 m in from the lip (called the inland position), but some factors such as soil salinity and plant cover were sampled every 20 m along the entire transect. Data that we will present for these two long transects will be an average of the two. A diagram of this "average" long transect is shown in Fig. 2.8.

Gradients in the Physical Environment

TEMPERATURE

As discussed earlier, oceans have a moderating effect on temperature extremes of nearby land areas. One can document the effects of this phenomenon even over very short distances in the grassland. Temperatures can be recorded over a period

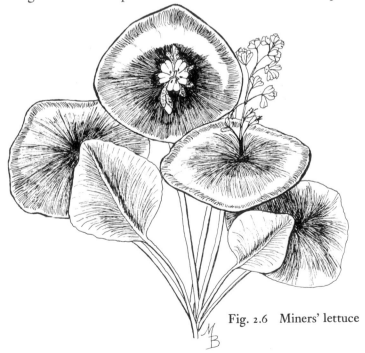

Fig. 2.6   Miners' lettuce

of time rather inexpensively with maximum-minimum thermometers (Appendix B). These thermometers not only indicate the current temperature on their dial faces, but they show
the maximum and minimum temperatures since the dial was
last checked. If the thermometers are checked once a day or
once a week, the maximum and minimum recorded must have
been sensed within that period. Some have a temperature sensitive probe at the end of a long lead, so that the probe may
be placed at specific locations above or below the soil surface.
We have placed such thermometers along both the short and
long transects to document temperature gradients.

For example, Fig. 2.9 shows the progression of daily summer maximum and minimum air temperatures 1 m above the
ground at the lip and inland positions on long transects. Differences were slight in minimum temperatures, but maximum
temperatures were much higher inland, averaging 66°F in
contrast to 62°F at the lip. The inland site also had a greater
daily range of temperature fluctuation (19°F) than did the
lip (15°F).

Temperature differences exist even over shorter distances. At noon on a clear August day, the following tem-

Fig. 2.7   Italian ryegrass

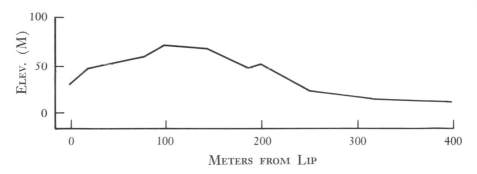

Fig. 2.8 Diagrammatic profile of elevation along the long grassland transect. Average of two nearly parallel transects.

peratures were recorded along the short transect 1 m above the ground: lip, 55°F; 10 m back, 60°F; 50 m back, 63°F.

But how important is air temperature 1 m above the ground to herbaceous plants only a tenth as high, with roots buried in the soil? Soil conducts heat poorly; consequently, the surface retains most of the heat energy that reaches it during the day and transfers only some of it to the soil just below or to the air just above. This results in a narrow, but hot, soil-air zone. For example, when air temperature on that August day mentioned above was 60°F 1 m above the ground, the soil surface was 100°F and soil 15 cm below the surface was 67°F. In warmer areas the soil surface can nearly boil an egg: on a summer day in the Nevada desert, air temperature 1 m above the ground may be 97°F and the soil surface 150°F. This big discrepancy between temperature near the ground, where the plants are, and temperature above it, where the Weather Bureau keeps its instruments, is one reason why Weather Bureau data often aren't what an ecologist really needs.

Maximum-minimum temperatures over the course of a year of the top 2 cm of soil are summarized in Fig. 2.10 for one location 20 m back from the lip. This is the seedbed where seeds may lie for several months before germinating. Notice that the range of temperature in topsoil is much greater than that in air 1 m above: compare the maximum-minimum fluctuation of 50°F in topsoil for late August with that of 15°F in the air (Fig. 2.9). With winter rains, topsoil temperatures fluctuate less and less until a period of near-constant temperatures, averaging about 55°F, is reached in late January.

Such a period of constant, moderate temperature, coupled with peak soil moisture, should be optimum for seed germination. The effect of temperature on germination can be determined in the laboratory by placing seeds in incubation chambers in which light, temperature, and humidity can be closely regulated. These chambers are quite costly, however, and a relatively inexpensive temperature gradient bar has proven it-

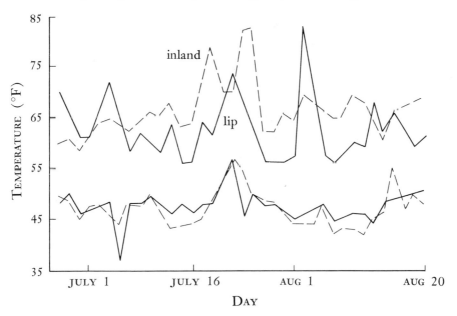

Fig. 2.9 Daily maximum and minimum air temperatures 1 m above the ground at two locations on the long transect. The lip position was 15 m in from the edge; the inland position was 310 m in from the edge.

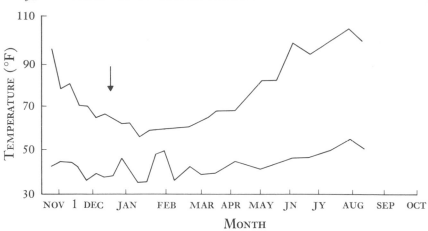

Fig. 2.10 Average weekly soil max-min temperatures 2 cm below the surface and 20 m in from the lip. Arrow shows time of peak germination at the site for most species.

self as a replacement. Very generally, the appartus consists of a slab of heat-conducting material, such as aluminum, which is heated at one end and cooled at the other, so that a continuous gradient of surface temperature results. Moist seeds in petri dishes can then be spaced out along the bar. A more detailed description is given in Appendix B.

Sea pink and the state flower California poppy, *Eschscholtzia california*, both grow near the point of the transect 20 m in from the lip: the seedbed temperatures for this point are shown in Fig 2.10. Using incubation chambers and the temperature gradient bar, the optimum temperature for germination of both is 57°F–65°F (Fig. 2.12 and Went 1970): very close to December–January seedbed temperatures.

But do sea pink and California poppy germinate in January at Bodega? No; like most grassland species, they do so in early December. Apparently, germination is not triggered by temperature as much as it is by the onset of winter rains, which occur in force in November. These winter rains dissolve the salt which has been deposited in topsoil by summer winds. As the rain water percolates through the soil it takes these salts with it, leaving the topsoil relatively salt-free. As will be discussed later, some grassland plants are inhibited from germinating by salts in the soil: germination of sea pink seeds, for example, is depressed to half its normal rate if the soil is even one-tenth as salty as sea water. Thus rainfall not only satisfies basic plant requirements for water as such, but the water affects plants indirectly through soil temperature and soil salinity.

SOIL TEXTURE

The quantity of rain is of course important, but the same amount can have a different effect depending on its distribution in time. Soil can store only a certain amount of water, and any additional precipitation in a storm simply runs off the surface and is lost to the plants. If all the rain fell in one torrent, some would be lost to runoff, but if it fell in showers over a longer time, all of it would enter the soil.

How much water a soil is capable of storing depends primarily on its texture—especially on the size of the soil particles of which it consists. Ultimately, all soil originates from parent rock which has been broken down by weather and organisms into smaller particles. Weathering may simply involve

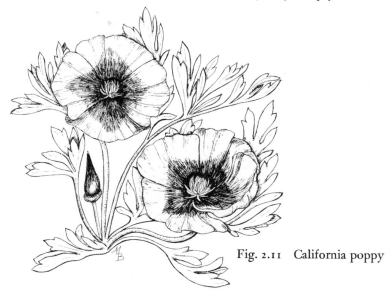

Fig. 2.11   California poppy

mechanical breaking of the parent rock. For instance, winds or waves may hurl sand against rock outcroppings and wear them down, or water collecting in pores of the rocks may expand at freezing temperatures to create fissures which later fragment the rock. Chemical weathering results in more profound changes in the mineral matter itself. Atmospheric gases, such as carbon dioxide and sulfur dioxide, become dissolved in rain water, producing acidic solutions that dissolve the rock. Plant roots secrete weak acids, and certain algae, bacteria, and lichens hasten the decomposition of some of the least resistant mineral matter.

Soil may form relatively rapidly if the type of parent material and type of climate are favorable. Conditions that hasten the rate of soil development are a warm, humid climate, forest vegetation, flat topography, and a parent material which is easily broken down. For example, Fort Kamenetz, built of limestone slabs in the Ukranian part of Russia, was abandoned in 1699. Today, a mature soil, 4 to 16 inches thick, has devel-

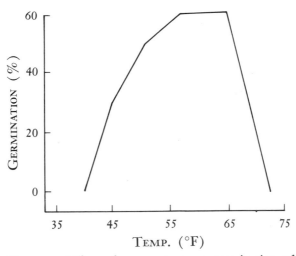

Fig. 2.12 Effect of temperature on germination of sea pink after fourteen days in darkness.

oped from the limestone. Conditions which retard soil development are a cold, dry climate, grass vegetation, steeply sloping topography, and parent rock material not easily broken down. It has taken from 1,000 to 10,000 years for mature soils to develop in northern areas scoured by the Wisconsin glaciation, which ended 10,000 years ago.

Soil scientists classify the particles which make up a mature soil into four size categories: gravel, all particles with a diameter greater than 2 millimeters (mm); sand, 2.0 to 0.02 mm; silt, 0.02 to 0.002 mm; clay, less than 0.002 mm. The microscopic clay particles are most important to the nutrient status and water-holding capacity of a soil. They are plate-like crystals mainly of aluminum, oxygen, and silicon compounds, which are put together in such a way that they carry a residual negative charge. Nutrient ions which are required by all plants for normal growth, such as calcium (Ca^{++}) or potassium (K^+) ions, are attracted by the negative charge and held to the clay particles. They are held tightly enough to prevent percolating rain water from dissolving them and leaching them from the root zone. Clay increases the water storage capacity of a soil because it increases the surface area, and water is held as a film over the surface of soil particles. Sand particles have 100 times the diameter of clay, and they fit together loosely, like so many marbles in a jar, so not only is surface area low, but water easily drains from the soil through the large spaces between the particles.

To realize the increase in surface area that comes with smaller size, consider one sand grain as if it were a cube 100 units long on each side. Its total surface area would then be calculated as follows: 6 sides x (area of each side, or 100 × 100) = 60,000 square units. Now suppose the sand grain is divided into clay particles, each a cube 1 unit long on each side. The area of each little cube is: 6 × (1 × 1) = 6 square units; but there are a total of 100 × 100 × 100 of them, so total surface area of all the clay is 6,000,000 square units. This gives 100 times the area of the sand.

Most soils contain mixtures of sand, silt and clay; the fraction of each in a soil can be determined in a number of ways (see Appendix B). The texture class of a soil depends on the fraction of each present, and a number of categories have been standardized by soil scientists. For example, a loam soil contains about 40% sand, 40% silt, and 20% clay. Other texture classes are mentioned in Appendix B.

The amount of sand, silt, and clay in topsoils along the short transect are summarized in Table 2.1. The proportions do not change greatly along the transect, and all can be called sandy loams. The amout of clay does more than double as one moves inland, but even at 50 meters it contributes relatively little to the total texture. Dune soil (Table 2.1) is nearly pure sand.

These grassland topsoils are very porous and crumble easily in the hand, and because of this they are easily compacted by foot or vehicular traffic, no matter how infrequent. With compaction comes decreased percolation of water and less aeration for plant roots, and the flora and fauna change dramatically in character—as will be discussed in Chapter 7.

Deer mice cause extensive erosion of this soil at the bluff, because their numerous tunnels collapse during storms or when walked on by heavy animals. Secondarily, the tunnels—

| | METERS INLAND FROM LIP | | | | | | |
	0	10	20	30	40	50	Dunes
Sand (%)	80.0	81.0	81.0	85.0	84.0	83.0	99.0
Silt (%)	18.7	16.5	16.9	11.7	13.0	13.5	0.5
Clay (%)	1.3	2.5	2.1	3.3	3.0	3.5	0.5

Table 2.1 Grassland soil texture (top 15 cm of soil) at different distances in from the lip. Dunes soil texture is given for comparison.

and those of other animals—increase soil aeration, which pro-
motes plant growth.

SOIL MOISTURE

The seasonal distribution of rainfall and fog was discussed in Chapter 1. Soil moisture closely follows the yearly
pattern of rainfall. Soil moisture can be measured by excavat-
ing about 100 grams (g) of soil, weighing it, drying it in an
oven, then weighing it again when dry. The percent moisture,
on a dry weight basis is:

$$\% = \frac{\text{wet weight} - \text{dry weight}}{\text{dry weight}} \times 100$$

Other methods of determining soil moisture are discussed in
Appendix B.

Fig. 2.13 summarizes monthly topsoil moisture 20 m from
the lip in level grassland. Soil mosture increased sharply with
the first fall rains in September and October, reached a peak
in January, and fell abruptly in March and April with the end
of the rainy season.

Just as temperature changes with soil depth, so does moisture. The topsoil layer provides an insulating barrier to loss of
water from lower depths. In September throughout the grassland, moisture content of soil 40 cm below the surface ranged
between 110% and 200% of topsoil moisture.

Even on the same day and at the same depth, soil moisture can change over short distances. Along the level grassland
transect, for example, topsoil moisture increased inland, as
shown in Table 2.2, for one time on a July day. The increasing water content inland probably reflects increasing clay
content and increasing distance from drying winds at the lip.
Hillside soils are considerably drier in summer than those in
flats, possibly because more of the rain runs off the surface of

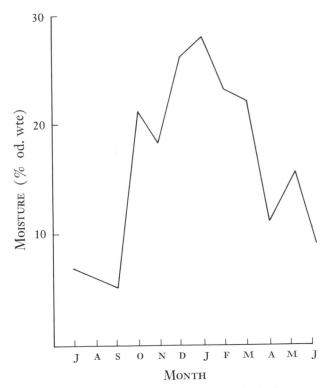

Fig. 2.13   Soil water content (dry weight basis, top 15 cm) 20 m inland from the bluff edge. Soil was sampled every three to five weeks.

hillsides and is not stored in the soil. Water in topsoil of flat areas during summer was 200% that of hillsides, and water content in subsoil of flat areas was 140% greater.

Fog provides additional moisture, but at Bodega Head it contains salt. Water collected in fog gauges during the summer had a salinity of 0.2% (or 2,000 parts per million). Although one feels wet when walking through the fog, low plants and the ground are dry to the touch except at the bluff edge and on the foredune near the beach. In the dunes, leaves and stems of beach grass may glisten with fog, and a ring of dark sand around the stem base indicates that moisture has run down and entered the soil. The incidence of fog is highest along the lip and hilltops, lowest in protected flats.

Another form of salt water intrusion occurs in winter rather than summer. Heavy surf, breaking upon rocks at the base of cliffs, often creates a thick layer of foam giving the appearance of soapsuds, which can be driven by high winds in great quantities across the level grassland. The foam, a normal feature and not the result of pollution, is conspicuous only between Horseshoe Cove and Mussel Point, where the shore is rocky and the bluffs are low (30 feet or less).

SALT SPRAY AND SOIL SALINITY

Soil salinity can be measured by determining the ability of a soil extract to conduct a current of electricity. Water alone is a poor conductor, but dissolved salts increase its conductivity. Salts in a soil sample are extracted, their conductivity measured, and concentration of salt calculated by standard procedures described in Appendix B. Concentration can be expressed as percent or as parts per million: 1000 g of salt in 1,000,000 g of soil or water equals a salt concentration of 0.1% or 1000 ppm. Sea water contains about 3.5% salt, but most land crops die if soil salinity climbs above 0.5%.

Meters from the lip	% Moisture
0	5.9
10	7.3
20	8.8
30	12.7
40	11.3
50	10.8

Table 2.2   Grassland soil moisture (dry weight basis, top 15 cm) at different distances from the lip on July 9, 1970.

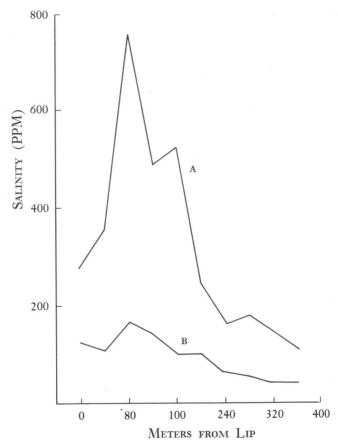

Fig. 2.14   Soil salinity (5–10 cm below the surface) on the long transect. A: average salinity of samples collected July 10, August 10, November 15, and January 18, 1969–70; B: salinity of samples collected 1 April 1970.

Soil salinity on the Head comes from salt spray and foam blown inland, and the amount of soil salt correlates well with the distance inland or degree of protection from wind. The summer pattern of soil salinity on the long transect, as sampled 5 to 10 cm below the surface every 20 m, is shown in Fig. 2.14 (curve A). By referring back to Fig. 2.8, it is immediately apparent that topography and soil salinity are correlated. At the ocean-facing lip salinity is moderate, about 300 ppm; it rises to peaks of 750 and 525 ppm on the hilltop and its shoulder, at approximately 100 and 200 m from the lip; it drops steeply along the lee side of the hill; and it reaches a low of 150 ppm in the protected flat area 350 m from the lip. A small increase in salinity about 270 m from the lip is probably due to spray emanating from Horseshoe Cove, for the transect passes closest to the cove at about that point.

The peaks of salinity actually correspond more closely to rock outcrops near the hilltop than they do to absolute elevation. Salinity is high near the base of these granitic rocks— as much as twice the salinity only 1 m away, and higher on the ocean-facing side. Possibly the surface of the rocks collects fog droplets and channels the saline water into the soil about their bases.

Soil salinity follows a seasonal progression. In late summer it is at its highest, from additions by fog and salt spray; in winter, leached by rain, it decreases; in early spring it reaches its lowest level; and in summer it increases again. Apparently, considerable rain must fall before the soil 5 to 10 cm below the surface becomes significantly leached, for we have found little difference between soil salinity on July 10 and January 18, even though rains begin in October. The A curve in Fig. 2.14 represents average salinity for samples collected four times between July 10 and January 18; the B curve shows soil salinity on April 1, after almost all rain had fallen. Salinity in April averaged half that of the "dry" season, but it still reflected topography, with small peaks at 80 and 200 m and a decline along the lee side of the hill.

Soil salinity changes over short distances in more complicated patterns than that just discussed above. As shown in Fig. 2.15, the *direction* of soil salinity gradients may shift with the season: in summer, salinity decreased inland, but in fall there was an intermediate situation. Salinity gradients also ex-

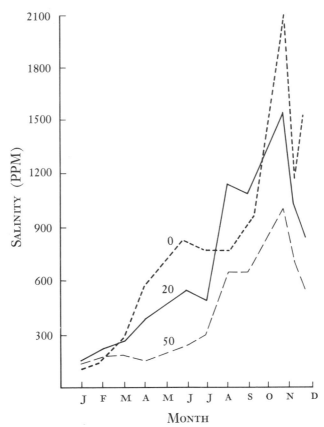

Fig. 2.15   Soil salinity (0–15 cm below the surface) at different distances in from the lip: 0, 20, and 50 m.

panded and contracted during the year: in January, salinity changed less than 100 ppm from the lip back 50 m (a difference below the sensitivity of most plants), but in June salinity changed along that distance by 600 ppm.

Shifts in salinity gradients during the year may be responsible for shifts in species distribution that occur during the same time. If a plant were sensitive to salinity, it might find germination and growth near the lip possible in January, but impossible in June—so its range would contract inland.

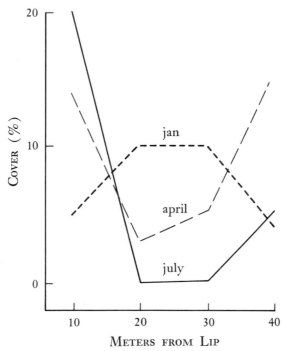

Fig. 2.16   Amount of ground covered by California poppy at four distances in from the bluff lip during January, April, and July.

Or, if it were salt-tolerant but a poor competitor with plants of other species, it would contract its range toward the lip during the growing season as competition lessens at the lip but increases inland.

Such a shift in zonation, possibly caused by soil salinity, was shown by California poppy. We measured its abundance at positions 10, 20, 30, and 40 m in from the lip by estimating the amount of ground it covered in sample areas 1 square m in area (see Quadrat Sampling Methods, Appendix B). Fig. 2.16 shows that in January poppy was most abundant 20 to 30 m back from the lip, but by July it was most abundant 10 m from the lip and was almost absent elsewhere. The amount of ground covered is a good indication of species vigor.

We mentioned earlier that germination of lip plants such as sea pink peaked in winter, and that this peak correlated in a general way with low temperatures and the onset of winter rains. From Fig. 2.15, it also appears that germination correlates with low soil salinity, for lip salinity is lowest in December and January. To test this correlation in the laboratory, we placed seeds of sea pink in petri dishes on a layer of sand saturated with different dilutions of sea water and kept them in the dark at 65°F for three weeks (see Appendix B for details).

When germinations were tallied (Fig. 2.17), it was clear that high salinity depressed germination. Germination was high so long as salinity was below 300 ppm. It is interesting to see, from Fig. 2.14, that soil salinity at the lip is below 300 ppm from January through March, the period of peak germination for sea pink (see also Fig. 2.12). Note that one correlation of rising soil salinity at the lip with time is that older sea pink plants which remain at the lip must be considerably more tolerant of salt than seedlings—but we have yet to document this in the laboratory.

Most plant physiologists believe that salts in the soil affect plants by making water less available, rather than by causing any toxic reaction by specific ions. As any solute, whether it be salt or sugar, is added to water, the ability of that water to move or to do work is reduced—this ability of water is called its "potential." Pure water is arbitrarily assigned a water potential of zero. Water with dissolved substances has a water potential less than zero, and the greater the concentration, the farther below zero the water potential falls, on the (negative) scale—usually measured in units called atmospheres (atm). Soil water below −15 atm causes most plants to wilt and eventually die, unless more water is added.

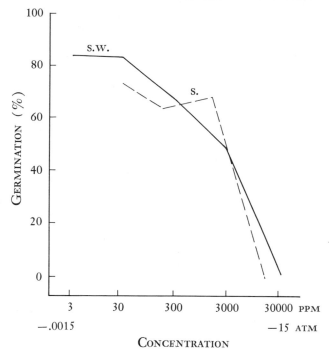

Fig. 2.17   The effect of increasing salinity (in parts per million salt) or of decreasing water potential (in negative atmospheres) on germination of sea pink. Germination noted after twenty-one days in darkness at 65°F.

One can show that salt affects sea pink through water potential rather than through toxic action by comparing the effect of sugar solutions on germination with the effect of salt solutions. As shown in Fig. 2.17, the effects of salt and sugar were identical, so a toxic effect can be ruled out.

If germination takes place at or very near the soil surface, as is the case for sea pink, then the seed may be exposed to even higher salinities than those shown in Fig. 2.15, which are for the entire top 15 cm of soil. Salinity can be as much as three times as high in the top 2 cm, due to recent additions by salt spray and evaporation of water which brings salt to the surface (Table 2.3).

| | METERS INLAND FROM LIP | | | | | | |
	0	*10*	*20*	*30*	*40*	*50*	*Strand*
Total salts							
top 3 cm	1,920	1,600	1,856	1,344	640	1,920	1,755
top 15 cm	832	1,088	1,152	896	896	634	563
Ions, top 3 cm							
calcium	44	80	66	22	18	30	
magnesium	103	103	101	36	36	38	
chorine	1,120	1,015	707	245	175	240	
sodium	62	51	37	25	21	21	

Table 2.3   Grassland total soil salinity, and contribution of four individual ions to that salinity (all in ppm). Total salinity was measured at two depths, ions only at one depth. Strand soil salinity is presented for comparison. August.

One should also keep in mind that we determined salt content from a 1:1 soil:water extract; that is, for every gram of soil, we added a gram of water in which to extract the salt —which is equivalent to a soil with 100% moisture (dry weight basis; see also Appendix B). Soil might contain this amount of water only during a heavy rain, but then rapidly lose most of it by drainage. Suppose it retained only one-fourth of the water: if all the salts which were in solution when the soil was saturated are still in solution, the salt concentration is four times what it was. Soil salinity on the lip in June was about 800 ppm as determined by 1:1 dilution (100% moisture), but soil moisture at that time was only 10%, not 100% (see Fig. 2.13), so the actual salinity sensed by a root would have been 8,000 ppm!

Finally, air-borne salinity may be just as important to species distribution as soil-borne salinity. Steep gradients in the intensity of salt spray exist in the grassland, just as they do in the dunes (see Chapter 4). Six salt spray traps (petri dish type, described in Appendix B) were placed every 10 m along the short transect and left in place for eight days in July. Table 2.4 shows that salt spray decreased inland. Five times as much salt was deposited near the lip as was deposited only 50 m inland. Lupine shrubs begin to increase further inland than 50 m, so they are exposed to even lower amounts than the 0.01 milligram/cm^2/day level at 50 m. Compare this to salt spray level of 1.0 mg/cm^2/day which we have measured on the beach: a difference of a hundred-fold.

There are no species in common between the beach and the lupine area. Is salt spray intensity the reason for the difference? We have compared the effect of salt spray on lupine and on sea rocket, a succulent herb of the beach. In the laboratory, three lots of seedlings of sea rocket and lupine were grown from seed in trays of sand kept continuously wet with tap water, and maintained in a growth chamber kept at 70°F. When the seedlings were about two weeks old, each lot was sprayed with sea water at different frequencies. By spraying patches of cheesecloth at the same time, it was possible to determine the amount of salt applied per square cm (see Appendix B). The treatments continued for two weeks, then average plant weight and development were noted.

Meters from the lip	mg salt/cm^2/day
0	0.04
10	0.05
20	0.04
30	0.03
40	0.02
50	0.01

Table 2.4 Deposition of salt spray (as determined by the petri dish method) at different distances inland from the lip, level grassland. Dishes were left exposed for an eight-day period in July, 1970.

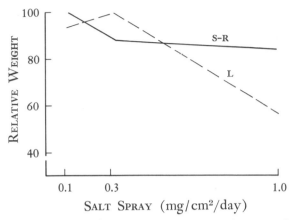

Fig. 2.18   The effect of salt spray on relative growth
of sea rocket seedlings (SR solid line) and on lupine
seedlings (L line). Seedlings were weighed after a
two-week spray program.

Fig. 2.18 shows that sea rocket seedlings grew well even
at the highest spray level, but lupine seedling growth was
considerably reduced at the highest spray level. The level of
salt spray on the beach may be 1 mg/cm²/day on a calm day,
but it is much higher on a windy day. Extrapolation of the
curves in Fig. 2.18 to higher spray levels indicates that lupine
seedlings would do even worse on the beach during winter
storms.

SOIL PH; ACIDITY AND ALKALINITY

The acidity or alkalinity of water, soil, or any mixture is
indicated by its "pH," a concept we will briefly review for
the general reader. Neutral solutions, in which the number of
hydrogen ions exactly balances the number of hydroxide ions,
have a pH of 7.0. Acidic solutions, which have an excess of
hydrogen ions, have a pH less than 7.0—the closer to 1.0, the
more acidic the solution. Basic, or alkaline solutions, which
have fewer hydrogen than hydroxide ions, have a pH greater

than 7.0—the closer to 14.0, the more basic the solution. The scale here is a logarithmic one; that is to say, each unit on the scale denotes a ten-fold difference in the number of hydrogen ions.

Although the theoretical range of pH is 1.0 to 14.0, soil pH typically ranges only from 5 to 8. Acidic soils, such as those beneath northern conifer forests, range from 3 to 5, agricultural soils range from slightly acid to neutral or 5 to 7, and desert or salt marsh soils range from 7 to 9 (sometimes as high as 11). Some plants have a narrow tolerance to range of pH. For example, azaleas grow well between pH 5 and 6, but tomatoes grow well only near neutrality. Soil pH affects plants mainly indirectly, by changing the solubility of necessary nutrient ions. Nitrogen is most soluble, hence able to be absorbed by roots, near pH 7, but it is relatively insoluble, and therefore unavailable to plants, at more acidic or basic pH; aluminum becomes so soluble at acidic pH that it may prove toxic to plants; calcium is most available in basic solutions and so on.

Soil pH can be measured in a number of ways (see Appendix B); using a laboratory pH meter, we have found that grassland soil pH is relatively constant throughout the year, and from place to place. Average pH of topsoil (0–15 cm) along the level transect, for example, increased only 0.1 pH unit from July to September. Average readings for July, August, and September are presented in Table 2.5, and one can see there is not a strong change in pH as one moves inland from the lip. Based on other measurements taken during the year, average grassland soil pH remains close to 6.5 (which is more acidic than dune soil, see Chapter 4). The uppermost 2 cm of soil averages 1.0 pH unit higher than soil just below it, putting seedbed conditions very close to neutrality. The more nearly neutral pH may be due to higher levels of organic matter or salt.

	Meters from the lip					
	0	*10*	*20*	*30*	*40*	*50*
Top 2 cm	6.7	6.6	6.4	6.5	6.6	6.9
Top 15 cm	5.7	5.6	5.6	5.6	5.7	5.5
Difference	−1.0	−1.0	−0.8	−0.9	−0.9	−1.4

Table 2.5   Grassland soil pH at two depths and at different distances in from the lip. Averages for July, August, and September.

Gradients in Plant Distribution

GRADIENTS OVER LONG DISTANCES

How well does the distribution of various plant species correlate with the environmental gradients just discussed? What is the exact distribution of these species? One method which allows for careful, objective examination of species distribution is a quantitative survey along a transect that cuts across many zones. Species are noted not only for their presence, but for their relative abundance or importance by measuring the amount of ground covered by them. Once a plant covers the ground with its foliage, it modifies the microenvironment—changing temperature and moisture near the surface and probably excluding certain other species. Basal leaves of some plants, for example, are especially effective in excluding nearby competitors. Hence, it seems reasonable that the area covered by each species should provide a good index of its importance in the community.

Obviously, the entire ground surface along a transect cannot be measured this way, for the time involved would be too great. Instead, small plots of ground ("quadrats") are sampled (see Appendix B). We employed an aluminum hoop with an area of one square meter, and placed it along the transect every 20 m. Dowels were hammered into the ground at these points, and a hole in the hoop crosspiece was arranged to just fit over the dowel; in this way, the same sample area could be relocated exactly at different times of the year.

When the grassland along the long transect was sampled in this fashion in July of 1969, a total of twenty-three species were recorded. This was probably only half of all the species present in the grassland at that time, but it included the important and abundant (dominant) species and excluded only the rare ones. Some of the twenty-three species are shown in a bar graph (Fig. 2.19): the thickness of the bar indicates the percentage of ground covered. The graph shows that few

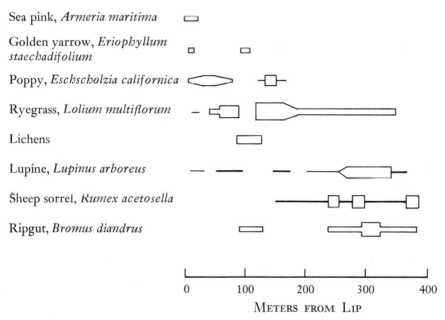

Fig. 2.19 Relative amount of ground covered by selected species on the long transect. The thickness of the bar indicates percent ground cover. For comparative purposes, Italian ryegrass covered about 27% of the ground 150 m back from the lip.

species were narrowly restricted in distribution. True, the herbaceous perennial sea pink was found only on the lip, and the low shrub golden yarrow was found only at the lip and around rock outcrops near hilltops, and several lichens were were restricted to rock outcrops near hilltops; but other species were merely more abundant in one area and were present over much of the entire transect.

Lupine shrubs occurred in virtually the entire 400 meters, but were most abundant in the flat inland portion. The lupine shrubs found closest to the cliff edge, back to the descending slope of the hill (0 to 180 m) are prostrate and blue to white flowered. Some consider them to be a different species (*Lu pinus varicolor*) although they seem to interbreed in the grassland. California poppy was most abundant at and behind the lip, but was also found on hilltops and in protected flats. Acid-tasting sheep sorrel, *Rumex actocella*, and the introduced grass ripgut, *Bromus diandrus*, were found on hillsides and hilltops but were most abundant in protected flats. In short, distinct boundaries in species composition, from one end of the transect to the other, could not be found. All it is possible to say is that a number of species reach peak abundance together in particular zones. The species which peak together can be called one another's "associates."

Common associates of lupine in protected flats, for example, include: bracken fern, *Pteridium aquilinum;* the vine-like, aggressive manroot, *Marah fabaceous;* the annual scarlet pimpernel, *Anagallis arvensis;* an occasional mushroom; and several thistles: sow thistle, *Sonchus asper*, bull thistle, *Cirsium vulgare*, milk thistle, *Silybum marianum*, and the ground-hugging *Cirsium andrewsii*.

Associated with the lichen-covered rocks on hilltops are: delicate blue dicks, *Brodiaea pulchella;* wild onion, *Allium dichlamydeum;* the spectacular wild iris, *Iris douglasiana;* soap plant, *Chlorogalum pomeridianum;* the succulent live forever,

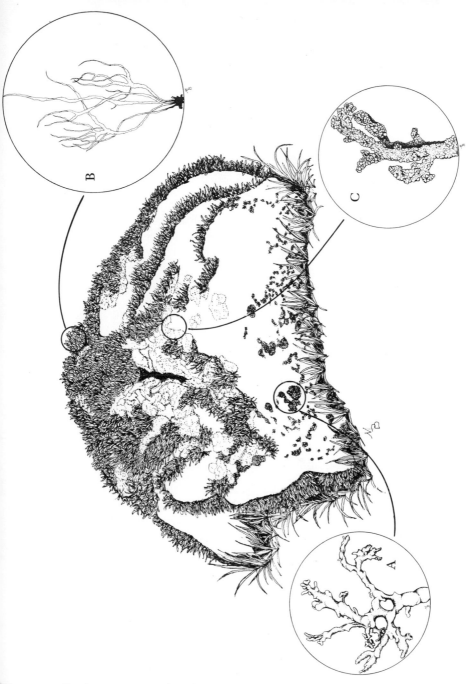

Fig. 2.20 Rock outcrop and lichens. 220a: small and red-brown *Caloplaca coralloides*. 220b: large and dark-green *Ramalina homalea*. 220c: pale green *Lecanora bolanderia*.

Fig. 2.21 Wild iris

Dudleya farinosa; and the obnoxious poison oak, *Rhus diver-siloba.*

Close behind the lip occur a number of species with relatively wide distributions but which seem particularly abundant here. Many of these flower in spring and early summer: goldfields, *Lasthenia chrysostoma;* cream cups, *Platystemon californicus;* baby blue eyes, *Nemophila menziesii;* and the yellow tidy tips, *Layia platyglossa.* Other perennials include seaside daisy, *Erigeron glaucus;* prostrate yellow mats, *Sanicula arctopoides;* the brittle, silvery beach sagewort, *Artemisia pycnocephala;* wild buckwheat, *Eriogonum latifolium;* the large carrot relative, *Angelica hendersonii;* gumweed, with sticky flower heads, *Grindelia stricta* ssp. *venulosa;* and a small shrub with pale, velvety leaves, *Phacelia californica.*

Another group of species are so widespread that they cannot be assigned to any one zone. All of the introduced grasses fall under this category: farmer's foxtail, *Hordeum leporinum;* soft chess, *Bromus mollis* (Fig. 2.22); hair grass, *Aira caryophylla;* ripgut; and Italian ryegrass, among others. Some native, herbaceous perennials are also widespread: buttercup, hedgenettle (which isn't prickly despite its name), *Stachys rigida,* and biennial cudweed, *Gnaphalium chilense,* are examples.

GRADIENTS OVER SHORT DISTANCES

Species distribution can also be treated mathematically. The botanist Jaccard thought that the degree of similarity between any two communities, A and B, could be expressed with the following formula:

$$\text{coefficient of community} = \frac{c}{a + b - c} \times 100$$

where a is the total number of species in community A, b the total number in community B, and c the number of species

Fig. 2.22   Soft chess

in common. The coefficient of community (CC) ranges from
100 for identical communities, to 0 for total dissimilarity.

We have used this formula to determine how species
composition changes over very short distances inland from
the lip. The "communities" to be compared (more strictly,
the groups of co-occurring species) were located every 10 m
along the short transect. A species list was compiled for each
community from quadrats (of area 1 m²) placed at the lip
and 10, 20, 30, 40, and 50 m inland. The quadrats were located on five parallel transects, so each position was sampled
five times. Every possible pairing of the six communities was
made, and its CC computed. The communities were then
ranked together in order of similarity on a dendogram (Fig.
2.23; see also Appendix B). The communities were lettered
consecutively with distance from the lip, so community A is
the lip, B is 10 m in, etc. The dendrogram shows that communities C and F had the greatest degree of similarity of any
pair, with a CC of 85. Communities D and E were next most
similar, with a CC of 75. More important, all the inland com-

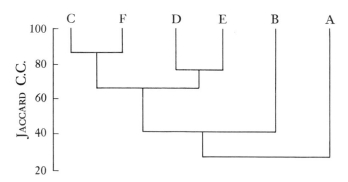

Fig. 2.23 Dendrogram showing relationship in species
composition among six communities in level grassland. The
communities are located on the lip (A), and every 10 m
inland to 50 m inland (F). Degree of relationship is indicated by Jaccard's coefficient of community (CC).

munities (C, D, E, and F) were closer to one another than they were to the lip communities (A and B). These communities had a cumulative CC of 65 among themselves, but had a CC of only 40 with B. And B was even less like A, for the two had a CC of only 25. One can conclude from the dendrogram that species composition changes very rapidly in the first 20–30 m from the lip; beyond that distance, however, species composition changes very little.

Jaccard's coefficient only takes two communities at a time into consideration. Actually, it may be more revealing to note loss and gain of species as one moves through a whole chain of communities. Pearson's coefficient does just that. The formula used is much more complex than Jaccard's (see Ap-

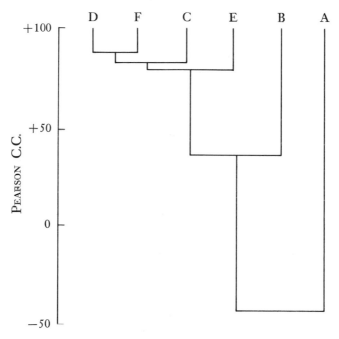

Fig. 2.24 Dendrogram for the same six communities shown in Fig. 2.23, but based on Pearson's coefficient of community.

pendix B), but the results can be interpreted in the same manner: the higher the coefficient, the more similar the communities; the lower the coefficient, the more dissimilar. Here, the maximum value is +100, minimum −100.

When the same six communities were analyzed with Pearson's coefficient, the dendogram in Fig. 2.24 resulted. Here, the close relationship between the four inland communities (C, D, E, and F) becomes even more obvious, and the uniqueness of the lip community is accentuated. There appear to be three types of communities along the short transect: one at the lip, an intermediate one 10 m back, and a homogeneous one beginning 20 m back and probably extending far into the grassland.

GRADIENTS OVER TIME

Many low herbs complete their cycle of growth early in the year, when moisture is optimum and before the grasses have overtopped and shaded them. These include miner's lettuce, *Montia perfoliata*, rock cress, *Arabis blepharophylla*, baby blue eyes, *Nemophila menziesii*, hedgenettle, *Stachy rigida*, and scarlet pimpernel, *Anagallis arvensis*.

Another group of species reach a peak of abundance in May, and these have mainly yellow flowers: cream cups, *Platystemon californicus*, goldfields, *Lasthenia chrysostoma*, buttercup, *Ranunculus californicus*, and tidy tips, *Layia platyglossa*.

In midsummer the grasses finish their vegetative growth and flower in profusion. Most of these are introduced annuals. The native perennial grasses, blue grass, *Poa scabrella*, and California brome, *Bromus carinatus* (Fig. 2.25), are uncommon today, but they may at one time have dominated this coastal grassland. Grazing of domestic livestock, which went on at Bodega Head for many years, opens the grassland surface and allows introduced annuals to invade, at the same time

Fig. 2.25   California brome

forcing perennials to rely more and more on germination, rather than vegetative growth, to maintain themselves each year. Grant Harris (1967) in Washington has shown for another grassland that introduced annuals may have faster root growth than do native perennials, and that they cause decline in growth of the natives by exhausting soil moisture ahead of the slower growing perennials. Possibly a similar phenomenon has tended to exclude native grasses on the Head. Harris' experiments are discussed in more detail in Chapter 7.

Lupine shrubs also reach their flowering peak in summer; flowers may be white, yellow, or blue even on the same flower stalk. Coast tarweed, *Madia sativa,* at this season dominates the aspect of hillsides with its tall, sticky shoots and small, white flowers.

In fall, several thistles reach full height and produce abundant flower heads: bull thistle, *Cirsium vulgare,* sow thistle, *Sonchus asper,* and milk thistle, *Silybum marianum.* A few fall-blossoming species are concluding a flowering period which they began seven to eight months earlier—a surprisingly long period, considering the changes in soil moisture, temperature, salinity, and other factors which occur during that time. California poppy, for example, produces flowers

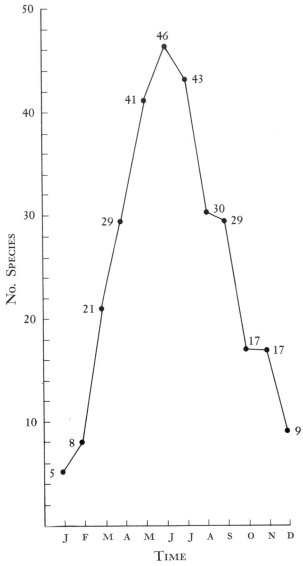

Fig. 2.26 Number of grassland species flowering each month of the year.

without a noticeable peak from March through October; baby blue eyes flowers from March through October; both sow thistle species peak in flowering in fall, but produce flowers from April through September. Flowering shoots of sow thistle might be even more obvious during the fall were it not for the deer. A herd of black-tailed deer, *Odocoileus hemionus*, graze very selectively on grassland species, and they feed heavily on sow thistle. Sometimes it is impossible to find any flower stalks remaining within a given area.

A prolific show of yellow flowers by a few species in May makes it appear that the peak flowering time is then; but a count of all species shows that the peak actually occurs in June (Fig. 2.26). The summer flowering peak contrasts with germination peak in December. What set of environmental conditions trigger germination at one season, growth at another, flowering at a third? Considering the number of yearly environmental changes examined in this chapter, it is not easy to pinpoint any one factor as being most important.

We can guess, however, that different influences trigger different stages in the life cycle, as the following experiment indicates. On February 21, plots of ground 20 cm on a side were cleared of natural vegetation. The plots were located every 10 m back from the lip in level grassland, to 50 m. In each plot, 100 seeds of sea pink were sown. Germination was noted three weeks later, on March 14, and seedling survival was noted nine weeks later, on April 26. The results (Fig. 2.27) indicate that conditions for germination were equally favorable over the entire transect, but conditions for seedling establishment improved with distance from the lip. A second series of plots, parallel with these, were prepared on December 15 and five mature sea pink plants were transplanted to each. The plants were completely vegetative at that time. Periodically, the plants were checked for growth and flowering.

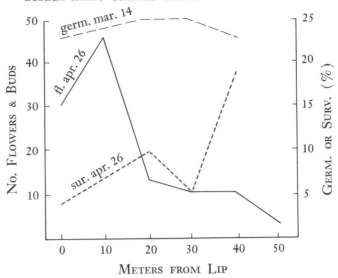

Fig. 2.27 Germination, seedling survival, and number of flowers and flower buds produced by sea pink in transplant gardens. Germination was noted on March 14, survival and flowering on April 26.

Fig. 2.27 shows the reproductive condition on April 26, about four and a half months after being transplanted from the lip. Many flowers and floral buds were produced in plants at the lip and 10 m back, and there was reduced production further inland.

Sea pink normally grows only at the lip; how do these results explain that restriction? Notice that if one looks at each stage separately, the "best" zone for sea pink shifts: for germination, all zones are equal; for seedling establishment, inland is best; for reproduction, near the lip is best; and if we had a response curve for vegetative growth, it might show yet a fourth pattern.

The point of this experiment is that this species occurs in a compromise location, a location which is not necessarily optimal for growth, reproduction, establishment, or germination, but a location in which all these processes may occur. Sea pink compromises between tolerance ranges throughout its life cycle and the environment of a number of possible habitats, but can only go through the entire cycle where the necessary conditions are all met. We shall see a similar pattern of response for a plant of the beaches, sea rocket, in Chapter 4. Clearly, if ecologists are to understand why species occur where they do, they must examine every stage of development.

Species distribution changes not only with the season, but with the years. Since the start of the nineteenth century, Bodega Head has experienced extensive human use, such as for farming and grazing. Manifestly, the vegetation on the Head now is not what the Indians knew several hundred years ago. And now that the area is a preserve, the vegetation of today will not be that seen a hundred years in the future. Aerial photographs taken in 1953 and 1969 clearly show that lupine shrubs have greatly expanded the area which they dominate. Lupine has increased on the sandy ridge where the dunes meet the grassland, in the protected flat area to the southwest of the laboratory, and on Mussel Point. The area dominated by lupine has increased three-fold in those sixteen years. If it were to continue to expand at that rate for another sixteen years, the entire grassland would be converted to a shrub vegetation type found along the coast north of Bodega Head. The California botanists Munz and Keck (1963) call that vegetation type northern coastal scrub and describe it as grassland dominated by the low shrubs lupine and coyote bush.

The components of the northern coastal scrub—as Munz and Keck list them—are all present at Bodega Head now, but they are strangely spread out. Lupine, golden yarrow, sagewort, and seaside daisy are in the grassland or at its lip; coyote bush is in the dunes (Chapter 4); and monkey flower, paintbrush, pearly everlasting, blackberry, and cow parsnip are in wet, protected gulleys (Chapter 6). Perhaps in time these components will merge at Bodega Head to produce a vegetation type that mirrors the coastal scrub of northern California.

What caused the rapid expansion of lupine? Chapter 7 hints that removal of grazing animals may have been responsible. If this is so, then the natural (undisturbed) community for the area may indeed be coastal scrub, and the grassland should progress to it through a succession of changes. These successional communities, for a variety of reasons, fail to maintain themselves, but the coastal scrub—the final ("climax") community—will maintain itself.

In theory, every environment is able to support a climax community, one with a characteristic composition, from region to region. The climax vegetation along the northern California coast is scrub; across Canada, coniferous forest; in the southeastern United States, deciduous forest. But all share the characteristic of being able to reconstitute and then maintain themselves following disturbance (whether by logging, burning, plowing, grazing, or whatever). Mixed hardwood forests in North Carolina, for example, if completely removed for cropland, are able to reconstitute themselves within two hundred years of cropland abandonment (Oosting 1942). The concept of climax communities was taken to its extreme by the American ecologist Frederick Clements, who, along lines of reasoning suggested in the Introduction, equated them with the adult stages of individual organisms. In 1916 he wrote that a climax community ". . . arises, grows, matures, and dies. Furthermore, each . . . is able to reproduce itself, repeating with essential fidelity the stages of its development . . . comparable in its chief features with the life-history of an individual plant."

Many ecologists disagree with Clements' definition of the climax as a "super-organism," but all agree that a succession of communities does follow disturbance. Why don't these communities maintain themselves? It may be because plants and animals modify their microenvironment—the soil and the atmospheric conditions near the ground. Light, humidity, soil pH, soil nutrient content, and wind are all affected. These modifications may make reproduction of the same species difficult. For example, in North Carolina, one of the successional communities is dominated by pine. Pine seedling survival is low in the shade beneath a closed pine canopy. Hardwood seedlings, however, can survive the shade, and the result is a gradual shift in forest composition from pine to hardwood. Early successional species—pioneer species—are quite different in their behavior from climax species. Pioneers do not depend on a few selected species; they have, instead, generalized feeding habits. Nor are they narrowly restricted to a certain type of soil or range of temperature; they may colonize widely different habitats. Whether plants or animals, they are relatively small, have a high reproductive capability, short life span, and great dispersal potential. Annual plants, flies, locusts, and many field rodents fit this description—an interesting collection of examples, in that these are also common weeds and pests. Climax species, on the other hand, are less tolerant of a range of habitats, are larger, have longer life spans, and reduced reproductive potential.

The Animal Community

The more familiar kinds of animals can move about, pursuing food and better living conditions. But their mobility does not mean that they occur at random. Like plants, they are restricted to certain zones of space and time at Bodega. Each adult bird or mammal has a home range in which the vital functions of food-getting, reproduction, and escape from predators are carried on. The size of such a home range depends on a number of factors, including how much food the animal needs, how much food is available in a given area, characteristics of the terrain, age, sex, size, and the density of the population. The home range occupied by a deer mouse, for example, is affected by the terrain and the territorial habits of these animals: they fight other members of their own species. In an open habitat such as grassland, it may be as much as 0.6 acre, but in more wooded habitats it drops to 0.2 acre. The lessened extent of the area occupied and defended seems to be associated with the presence of natural breaks or barriers in the terrain that prevent the mice from frequently contacting each other. In a laboratory cage, more mice can be happily accommodated if semi-partitions are placed within it. As another example, T. W. Schoener (1968) has shown that size of feeding range for birds depends both on body weight and type of diet (animals vs plant food). The red-tailed hawk, the marsh hawk, and the sparrow hawk are all carnivorous and share much the same diet at Bodega Head, but the area in which they feed increases with their body size: for the small sparrow hawk, which weighs about 100 g, this area is 350 acres; for the larger marsh hawk, weighing 500 g, over 600 acres; and for the red-tailed hawk, weighing 1100 g, 1000 acres. In contrast, the common grassland song sparrow, who includes both plant and animal matter in his diet and weighs 22 g, ranges over only half an acre. The reason for these relationships is an economical one. A big animal needs more food, hence, more area in which to find it. And since more plant material is available than animal, herbivores can make do with less space.

Information on small mammal abundance can be obtained with live traps. The two long grassland transects served as base lines for extensive mammal trapping. Every 20 m along the transects, a group of three traps were set. (If only one trap is used, the census is incomplete, because other animals will not be able to walk into the trap once it has been occupied and sprung.) Traps used for small animals such as shrews and mice were small, rectangular steel boxes called "Sherman" traps (10 inches long, 3 inches wide, 3 inches high). The trap, which we baited initially with peanut butter and rolled oats, is open at one end; when the animal walks in, he trips a release which closes the door. We found it unnecessary to rebait these traps; curiosity on the part of the animals was enough to encourage them to enter.

How far an animal has travelled can be determined by noting in which traps he is caught and recaught over a period of time. The first time he is caught, two of his eighteen "toes" are removed—a different two for each animal—and he is released. Later recaptures indicate how far he travels. The number of animals recaptured can also be used to calculate the size of the entire population, by making a number of assumptions and using standardized formulas (see Appendix B).

Larger animals (such as raccoons) were trapped with baited "Havahart" traps, placed in areas subjectively estimated as of likely animal activity such as water holes and well-used runs.

To illustrate the role that plants play in determining the composition of an animal community, we will organize a description of the community around one plant: lupine. Lupine shrubs are not distributed uniformly over the Head, nor do they enter into the food habits of all the grassland animals. However, they do dominate a part of the grassland, and they are critically important to a great number of animals. Indeed, most of the animals are ultimately linked to lupine shrubs through direct or indirect feeding relationships. Some of these animals are strict herbivores; others are omnivores, ready to eat plant or animal; and others are strict carnivores, only indirectly dependent on lupine.

HERBIVORES

Strict herbivores include a wide variety of animals: insect larvae, birds, gophers, rabbits, and deer. Some of these feed directly—even exclusively—on lupine, others feed on plants associated with lupine.

The wooly, salt marsh moth caterpillar, *Estigmene acraea*, the common black millipede, *Bollmaniulus fareiter*, and the yellow-banded larvae of the California tussock moth, *Hemerocampa vetusta*, all feed directly on lupine foliage. The salt marsh moth caterpillar, abundant in early summer, feeds primarily on the buds. His wooly brown body closely resembles the lupine pods, but we doubt there is any protective significance to this, because most of the larvae have already gone into pupation by the time the pods ripen in late summer. The black millipede is most common in spring and early summer, but is present throughout the growing season. The yellow tussock moth caterpillars are so abundant in late summer of some years that they seem to drip off the branches. Their numbers were so high in 1970 and 1971, that about 20% of all lupine shrubs were denuded of foliage. Many of the denuded shrubs were not killed as a result, however; new foliage was produced in fall. Another common insect herbivore of lupine is the larva of the painted lady butterfly, *Vanessia cardui*. These insect populations fluctuate widely from year to year. *Hemerocampa*, for example, was uncommon in the spring of 1970, abundant in 1971 and 1972, and rare in 1973. *Vanessia* was abundant in 1973. Other insects, such as the crane fly, *Tipula* sp., many occasionally suck plant juices. Ants have been seen carrying lupine seeds to their nests, and sometimes the seeds are still attached to the plant when collected. A variety of bees, flics, and beetles feed on pollen or nectar and count as herbivores, but also function in a symbiotic relationship with the plants resulting in flower pollination and eventual reproduction.

Many lupine shrubs die each year on the Head. The foliage first wilts, then turns grey and brittle, and finally falls off; and later the dead limbs are beaten to the ground by winter storms and are decayed by soil organisms. Sometimes an acre patch of many shrubs succumb at the same time, sometimes scattered dying shrubs appear next to perfectly healthy ones. Death of lupine shrubs has often been ascribed to grazing damage by the salt marsh moth caterpillar, but an examination of dying shrubs fails to show extensive grazing damage. Pathogenic fungi or bacteria are also absent. Stem galls around the eggs of a chalcid wasp that parasitizes the stems are sometimes present, but these are not numerous enough to cause plant death. Excavation of the roots, however, will show that they have been hollowed out by the larvae of the swift moth, *Hepialus behrensi*. The larvae are also responsible for death of lupine at several other localities along the coast, including Half Moon Bay, about 100 miles to the south.

The lined snail, *Helminthoglypta arrosa*, does not feed on lupine, but does feed on some plants associated with it. The snail is active mainly in the wet season; in the summer it becomes dormant ("aestivates"), withdrawing into its shell. In the northern part of the United States, and in its high mountains, the unfavorable period of time for animal activity is winter, with cold temperatures and snow or ice covering potential food sources—and there, hibernation, or winter rest, is common. At Bodega, however, winters are merely wet and cool, and plant growth is lush; the unfavorable season is the warm, dry summer—and several animals avoid it by aestivation. That the snail is avoiding drought, and not warm temperatures, is shown by its activity in the fresh-water marsh,

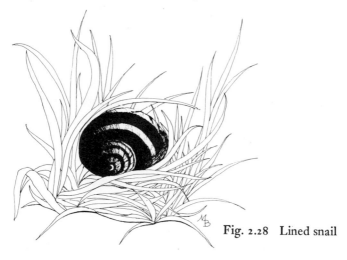

Fig. 2.28 Lined snail

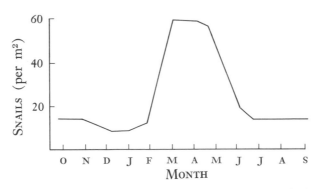

Fig. 2.29   Seasonal changes in population density of the lined snail. Courtesy of K. van der Laan.

which is continuously damp but no cooler than elsewhere on the Head: it does not aestivate in the marsh. The snail remains dormant in summer beneath lupine shrubs until October when heavy rains trigger resumed activity. As shown in Fig. 2.29, population density is low at this time—about fifteen snails in every square meter—but it drops even lower as the animals mate and begin to lay eggs in December. Population density skyrockets in February when eggs begin to hatch, and it remains high through April, despite predation by deer mice. The snails feed on vegetative matter, but apparently are very selective about food sources, according to a study made by Kenneth van der Laan. They reject bracken fern, sheep sorrel, lupine, grasses, hedgenettle, and manroot, all of which are prevalent in the lupine areas, but they do eat fiddleneck and wild heliotrope, two native annuals. It may be more than coincidental that the favored foods during the period of maximum snail activity are annual herbs whose maximum abundance occurs at the same time.

Predation by deer mice and other animals in spring reduces the population density from its peak of sixty per square meter to one-fourth that, and in June and July the lined snail once again aestivates—possibly triggered by increasing aridity. Although the snail imitates the cycle of an annual plant, it is actually a perennial organism. The time necessary to reach mature size, shape, and capacity to reproduce is about three years.

The California meadow vole, *Microtus californicus*, also utilizes herbaceous plants associated with lupine. His prodigious appetite can greatly affect the composition of the grassland, even though he is only active early in the morning and evening, and generally stays within established paths or runs. Currently, voles are less abundant on the Head than deer mice; but this may not always be the case, for their population density can fluctuate tremendously over a period of years. An intensive two-year trapping program by Charles Krebs (1966) in grassland near Berkeley, California (grassland with plant composition and climate similar to that at Bodega, 60 miles away), revealed that population density of this vole can fluctuate from 1 to 325 per acre from year to year. This is a magnitude of change greater than the famous "cyclical" plunges and peaks of lemming populations in the northern tundra of Canada, Scandinavia, and Alaska.

What causes such dramatic population changes? Biologists aren't sure. Krebs showed that in his grassland, the cause was not fluctuations in abundance of food, moisture, or any other physical factor. He kept track of vole population in several separate fields, each isolated from the other but all within the same small, climatic area. Populations rose and fell in different cycles in different fields: peak population in one field

Fig. 2.30   Meadow vole

occurred in September, at the end of the dry season, but peaks in two other fields occurred in March, at the end of the wet season. At one time, two fields, separated only by a gravel road, showed tremendous differences in population size—one had 300 voles per acre, the other had only 60. In one experimental area, he killed every pregnant vole that was captured —a total of nearly 2000 mice removed in a two-acre area— but the population density continued to grow as fast as that in the surrounding field. Apparently, once a population is for whatever reason headed for explosion, even an unnaturally high, artificially intensified rate of predation cannot keep the explosion down. And sometimes the explosion is not local, as it was in Krebs' study. K. F. Murray (1965) has reported details on the outbreak of another vole species—not found on Bodega Head—which occurred simultaneously throughout its northern California range in 1957–8.

During our study, vole population on the Head appeared to be rising: they were trapped five times more often in the summer of 1970 than in the summer of 1969. In years of low population density they favor the lee side of hills and protected flats with considerable grass and lupine cover, but in years of high density they undoubtedly spill into less favorable habitats, such as the wind-blown hilltops and cliff edges, where we did not find them. Based on investigations elsewhere (see King 1968), the home range of a vole is on the average much smaller than the 0.2 to 0.6 acre range of a deer mouse.

Pocket gophers, *Thomomys bottae*, are burrowing mammals which feed on roots and bulbs that intersect with their tunnels. They have strong, wide forepaws and large, chisel-

Fig. 2.31   Brush rabbit

Fig. 2.32   Black-tailed deer

like teeth; they collect food and temporarily store it in fur-lined cheek pouches. Gophers routinely leave their tunnels to forage above ground and become accessible to such predators as the barn owl.

Brush rabbits, *sylvilagus bachmani,* have most often been observed on Mussel Point among thick stands of lupine shrubs. The rabbits use lupine as protective cover, and they feed upon associated plant species such as brome, clover, fescue, filaree, and tarweed. They do not travel well through thick grass, and therefore prefer sparse ground cover beneath lupine shrubs, or else follow well-defined trails.

A herd of black-tailed deer roams the grassland, fresh-water marsh, dunes, and gulleys. The herd consists of at least twenty does (each with an average of one plus fawns each in summer) and three bucks. Adept at hiding among the lupine shrubs or in clumps of dune grass, they are seen far less often than would be expected, but are regularly encountered in the late evening. Deer are selective browsers, feeding on brome, buckwheat, and sow thistle. They forage primarily in the evening and early morning and rest under lupine in different areas of the Head during the day. A pond in the dunes, springs, and the fresh-water marsh provide water for them. Carcasses of fawns have been occasionally found, but cause of death could not be ascertained. Large predators, capable of killing deer, have only rarely been sighted in the area. One mountain lion, *Felis concolor*, was sighted near Mussel Point in 1970, several unconfirmed sightings have since been made.

The California state bird, the California quail, *Lophortyx californica*, concludes our list of strict herbivores. Quail are year-round residents of the Head. Seeds of lupine and of other plants form the bulk of their diet, but fresh leaves are also eaten in spring.

OMNIVORES

Some of the most common animals in the grassland are omnivores. The most abundant one of all is the deer mouse, *Peromyscus maniculatus*. This little, big-eared mouse feeds on the seeds of lupine and other plants, and on seasonally abundant invertebrates such as insects, the common black millipede, and the lined snail. From information gathered else-

Fig. 2.33   California quail

where (see Martin, Zim, and Nelson 1951), it is known that deer mice tend to shift their diet with the season. In winter they tend to concentrate on plant material, but in summer animal material makes up a large portion of the diet. In turn, it is a basic food for several predators: the marsh hawk by day, the barn owl, *Tyto alba*, and gray fox, *Urocyon cinereoargenteus*, by night. It therefore has very cryptic habits. Although its fur camouflages well with the vegetation, it still demands heavy plant cover; deer mice are most abundant beneath lupine shrubs in protected flats, and beneath sagewort and golden yarrow at the bluff edge. Deer mice are rare on hilltops, where vegetation is less than eighteen inches high.

The deer mouse population density, unlike that of voles, remains relatively constant from year to year. Apparently, reproduction—though prolific—is in balance with predation, disease, and other mortality. Their sometime preference for plant food, and other of their habits, bring them into close contact and potential competition with voles. However, minor adjustments in food habits and periods of activity lessen the competition. For example, the deer mouse is active later in the evening and earlier in the morning than the vole; the deer mouse concentrates on seeds, the vole on foliage, and even the choice of food plant species differs; and the deer mouse seems less restricted to runs than the vole.

Another mouse, the house mouse, *Mus musculus*, is rare on Bodega Head. It has been introduced to North America accidentally many times over the past 500 years from ships, and it is now a common household and agricultural pest in the United States. Because it evolved in European ecosystems, it is usually not compatible with the native fauna in undisturbed habitats, such as coastal grassland. Native species, like the deer mouse or vole, already occupy a niche much like its own, and these have not been ousted by the invaders. But in disturbed areas, where these native species are absent, the house mouse can reproduce to much higher densities. At Bodega, we have trapped the house mouse only in low grass near roads or buildings. For a strange story of the disappearing house mouse, see Chapter 7.

The most conspicuous bird in the grassland is the omnivorous white-crowned sparrow. Although year-long residents of the Head, they are most common in spring and summer. They have been observed in the lupine area and at the bluff edge, eating seeds of sea pink, California poppy, miner's lettuce, and tarweed, depending on the time of year. Grasshoppers and caterpillars make up a considerable portion of their diet in spring and summer. Like many other animals, this bird has two peaks of foraging activity: early morning and late afternoon. Many biologists feel that feeding behavior like that is triggered by temperature, but Martin Morton (1967) has shown it to be triggered by solar radiation and not temperature. For one thing, his field observations told him that morning and afternoon foraging peaks were most pronounced on clear days, rather than on cloudy ones, and that feeding activity in the morning increased as the sun came up and before air temperature began to rise. In the laboratory, he put a number of wild birds in special cages whose light intensity and temperature could be regulated: the results showed that foraging behavior could be artificially triggered by manipulating the light intensity independently of temperature.

Most of the small birds at Bodega Head are omnivorous, at least in part, but plants are heavily favored over insects as a food source. Common associates of the white-crowned sparrow are the savannah sparrow, *Passerculus sandwichensis*, and the song sparrow, *Melospiza melodia*; snails and other animals make up an extremely high percentage of their summer diet, but plant food predominates the rest of the year. Their eggs from nests on the ground in turn serve as food for other animals. The western bluebird, *Sialia mexicana*, and red-winged blackbird, *Agelaius phoenicius*, are occasionally seen. A few ravens, *Corvus corax*, frequently soar in updrafts along the grassland cliffs. Migratory birds seen in spring and summer include the colorful American goldfinch, *Spinus tristus*. The red-shafted flicker, *Colaptes cafer*, and meadow lark, *Sturnella neglecta*, in contrast, are common in winter and rare in summer.

A large, omnivorous mammal common throughout the grassland is the striped skunk, *Mephitis mephitis*, which eats a multitude of animal matter and the fruit of some plants. Animal food includes the Pacific tree frog in spring, deer mice all year, the common alligator lizard in spring and summer, and the eggs of ground-nesting birds. Skunks are active in the evening; during the day they remain in burrows of gophers or other animals which they appropriate and widen to suit themselves. Raccoons, *Procyon lotor*, are often thought of as strict carnivores, but in fall and winter, vegetable matter may make up 50% of their diet. Unlike the skunk, these animals range over a number of habitats and are not restricted to the grassland. They require a lot of food and water. Tree frogs, mice, an occasional small rabbit, alligator lizards, and garter snakes are eaten, and they sometimes create a modest distraction when raiding peoples' garbage cans.

CARNIVORES

The term "carnivore" tends to be associated with big animals like bears and lions, but size has nothing to do with it. Possibly the most voracious carnivore is the little vagrant shrew, *Sorex vagrans*. Shrews are classed as insectivores, and

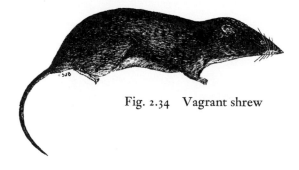

Fig. 2.34   Vagrant shrew

as the name implies they eat primarily insects. Shrews have a very high metabolic rate and eat constantly to sustain it: one has been observed in captivity to eat the entire fleshy part of a deer mouse and three millipedes in a twelve-hour period. This was far more than the animal's own body weight. On a weight-to-weight basis, this is like a man finishing off a 170-pound steak! Shrews are found mainly along the lee side of hills; they are rare at hilltops, absent at the cliff edge, and absent in the lupine area where deer mice are so abundant. It may be that the many deer mice dominate the available habitat space and thus exclude shrews, even though their food habits differ. Shrews are not a favorite food of mammalian and avian predators. They exude a skunk-like odor, and it is suspected that their flesh is bitter. The common garter snakes, *Thamnophys sirtalis*, however, do feed on shrews.

The bright green color and markings of the tree frogs, *Hyla regilla*, make them difficult to see in the grassland, and their foot pads allow them to climb in shrubs. Despite the name, they are rarely found on trees. Abundant during spring, they feed on insects. The alligator lizard, *Gerrhonotus coerleus*, is a hibernating species, prominent from May to November, but dormant in underground burrows from December through April. It feeds on millipedes and snails in summer. Another small carnivore is the garter snake, of which there

Fig. 2.35   Northern alligator lizard

are two species. It may occasionally eat a young mouse, but usually is more interested in bird eggs or insect larvae. The grassland is riddled with tunnels and holes, utilized by the alligator lizard, mice, shrews, voles, snakes, and skunks.

When the word "hawk" is mentioned, a picture of a lone bird circling slowly and quietly high above the ground may come to mind. But just as terrestrial plants and animals occupy different spatial zones, so do hawks. In this case, the zones are flight paths at different heights from the ground. The hawk with a low-level flight plan is the marsh hawk, *Circus cyaneus:* during the day any of the members of a single marsh hawk family on the Head can be seen gliding fast, three to four feet above the vegetation searching the ground for a careless vole or deer mouse. The male is present in spring and early summer but is gone through most of the summer; the female and usually two youngsters remain throughout the year. The youngsters appear nearly fully grown by August, but the mother remains very protective of them. The nest is on the ground usually in the tall beach grass of the dunes. The marsh hawks also circle the fresh-water marsh. The nocturnal equivalent of the marsh hawk is the barn owl, *Tyto alba*. There are at least five barn owls on the Head, but we do not know how many nests. Nests and daytime roosts are in trees protected from wind and sun in a gulley which faces the harbor.

Sparrow hawks, *Falco sparverius*, fly 30 to 40 feet off the ground, mainly over the short grass areas on hilltops and hillsides during the day. Keen eyesight and agility permit them to make sudden descents for deer mice, voles, or even large insects such as grasshoppers. Another daytime predator circles the grassland 60 feet or more from the ground: the red-tailed hawk, *Buteo jamaicensis*, which is occasionally seen floating for hours in updrafts over the hills. At the same altitude is the turkey vulture, *Cathartes aura*, but he is not a predator and has never been known to descend on live prey. He is a scavenger, one of the decomposers. Turkey vultures are seen at Bodega throughout the year.

Some of the larger predatory mammals have occasionally been seen at Bogeda Head, or else detected indirectly: gray fox, *Urocyon cinereoargenteus*; American badger, *Taxidea taxus*; long-tailed weasel, *Mustela frenata*; bobcat, *Lynx rufus*; feral cats, *Felis domesticus.* coyote, *Canis latrans; and* mountain lion, *Felis concolor.*

Gray foxes feed heavily on rabbits, and also prey on birds, mice, and other small mammals. In other areas, where summer and fall fruit are available, foxes are omnivores. The badger is adept at digging out underground burrows in search of mice and gophers with his powerful front paws; he is nocturnal and bad-tempered, fighting fiercely to defend himself when provoked. The long-tailed weasel is also a predator of burrowing animals; his slim body allows him to "weasel" his way into mouse and gopher holes without digging. Weasels also eat birds: one was observed in the grassland attacking a ring-necked pheasant far bigger than itself. Feral house cats, often several generations removed from domestication, are common mousers in the grassland.

The coyote and bobcat have been detected only by "signs"—fecal remains, called scats. The size, shape, and contents of the scat are very characteristic for certain animals, so that the presence of the animal can be told just as surely as from tracks. The mountain lion was seen only once, during the day at Mussel Point. If he is a regular visitor, the Bodega Head area is certainly on a small fraction of his home range which probably covers more than thirty square miles.

A GENERAL OVERVIEW

A portion of the food web of the grassland, incorporating several species which have been discussed, has been presented in diagrammatic form as a "food web" (Fig. 2.36). Here, one can see how different portions of the food resources derived from plants are utilized by animals. Roots of lupine are attacked by two organisms, buds by a different one; a fourth parasitizes stems; another eats seeds, and yet another foliage. Many of the organisms which utilize lupine directly have been omitted from this chart.

Lupine may also encourage growth of associated species, such as fiddleneck, by changing the microenvironment and making it more suitable. These associated species may have their own particular grazers; in the case of fiddleneck, it is the lined snail; in the case of sow thistle, black-tailed deer; and so on.

The herbivores are then attacked by carnivores, such as the meadow lark, the marsh hawk, and the gray fox. Dead remains of plants, herbivores, and carnivores alike are utilized by decomposers. Again, size has little to do with food habits, for the decomposers range from microscopic bacteria to the turkey vulture.

The chart in Fig. 2.36 gives just a bald outline of the interactions among organisms in the grassland. If one drew a food web indicating all species, there would be an extremely complicated network of interconnecting lines. Most animals would prey upon a diversity of species, and would in turn be preyed upon by a diversity of species. Such interconnections make the community resistant to great fluctuations in the abundance of any one organism. For example, if one species were to become excessively abundant, its many predators would begin making it a greater part of their diet and thus dampen the effect of the abundance by spreading it out in the food chain. Or, sudden scarcity of one species induces its predators to shift their diet to other prey. The number of links in a food chain is a measure of community stability.

Fig. 2.36 would give a more adequate impression of the dynamics of the system if it showed the absolute amount of

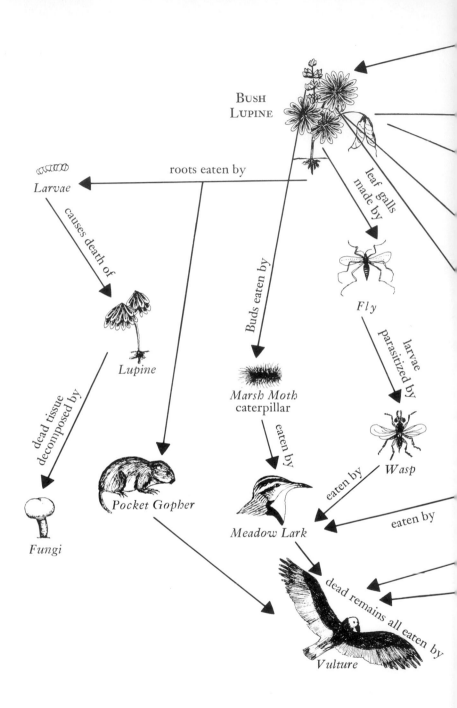

BUSH
LUPINE

roots eaten by

Larvae

causes death of

Lupine

leaf galls
made by

Buds eaten by

Fly

larvae
parasitized by

dead tissue
decomposed by

Marsh Moth
caterpillar

eaten by

Wasp

eaten by

Fungi

Pocket Gopher

Meadow Lark

eaten by

dead remains all eaten by

Vulture

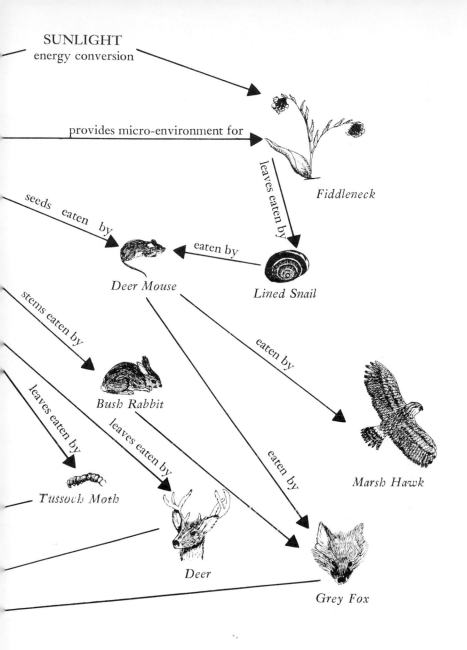

SUNLIGHT
energy conversion

provides micro-environment for

Fiddleneck

leaves eaten by

seeds eaten by

Deer Mouse

eaten by

Lined Snail

stems eaten by

leaves eaten by

Bush Rabbit

leaves eaten by

Tussock Moth

eaten by

Deer

eaten by

Marsh Hawk

Grey Fox

Fig. 2.36 A portion of the food web of the grassland ecosystem. Many organisms and interrelationships are not indicated.

energy that accompanies each arrow. How much of the lupine, for example, is utilized by the deer mouse, and how
many deer mice are required to keep one marsh hawk flying?
How does the efficiency of energy transfer here compare with
that of a coastal pasture or field of brussel sprouts? Caloric
values for other regions, as well as the laws of thermodynam-
ics, indicate that energy is lost each time it is transferred from
one organism to another; consequently, it takes a lot of grassland vegetation to support relatively few herbivores and even
fewer carnivores.

Plant Productivity and the Food Chain

The internal combustion engine, for all its nefarious by-
products and effects on human social behavior, is a wondrous
machine. All its parts are intricately and effectively fashioned
together so that the movement of one part is translated into
the movement of another. The pistons turn the crank shaft,
which propels the flywheel, which twists the drive shaft,
which cranks the rear differential gears, which spin the wheels,
which move the car, which hurtles us down the highway. But
the prime mover—that which starts everything going—is combustion within the piston chamber. Chemical energy in gaso-
line is converted there, by combustion, to mechanical energy.
So it is with the grassland ecosystem. The enormously complex food chain, of which we have produced only the rudest
sketch in the previous pages, includes hundreds of species of
organisms, which eat and are eaten, which parasitize, reproduce, nibble buds, drop like a rock from the sky on prey, or
decompose gophers that may die underground. But all this
would never occur were it not for the conversion of radiant
energy into food energy by green plants.

How much energy gets converted, and how efficiently
this process goes on, profoundly influences these animals, de-

termining the sizes of populations and their very presence on the Head. The conversion of energy by plants is termed "primary productivity." What is its magnitude and efficiency at Bodega Head?

A standard technique for measuring primary productivity involves clipping all vegetation down to the ground in a well-marked sample area, then clipping all plant material which grows up in its place at the end of a suitable period, usually a growing season or a complete year. At Bodega, we clipped all vegetation which had come up in the period from October to July. This represents the growing season—about nine months. Several plots, located at different distances inland from the lip, were clipped. The above-ground matter was dried and weighed, and root weight estimated (see Appendix B) and added.

Average primary productivity was 527 grams per square meter; it changed very little with distance from the lip. Expressed on a daily basis, that productivity is 1.8 $g/m^2/day$ for the 289 days. How does this compare with other areas in the world? Somewhat on the low side. It is within the expected range of 0.5 to 3.0 for temperate grasslands of the world (see Odum 1959), but it is below the 3.0 to 10.0 range for tropical grassland and forest, and even more below the 10.0 to 25.0 range for estuaries, coral reefs, and sugar cane fields. About all that it consistently exceeds is the deeper waters of the ocean (less than 1.0) and the desert (less than 0.5).

What is the food energy content of this plant matter? Using the data of others (see Appendix B), each gram of plant tissue in the grassland is equivalent to 4,300 calories, or 4.3 kcal. This means a total of 2,266 $kcal/m^2$ was produced in the course of nine months.

This does not represent all of primary productivity—that is, all the plant material produced during that time—because animals had been feeding on the plants and removing materials all that time. It would only be possible to know what all (net) productivity was, if the plots had been fenced and fumigated or otherwise protected from herbivores. So we shall have to confine our discussion to the partial net productivity. In work done elsewhere, in similar grasslands, it was found that herbivores utilized 10% of all plant matter produced and the rest fell to the decomposers. If that were true at Bodega Head, then total net productivity was 2,493 kcal/m^2 instead of 2,266 kcal/m^2.

It may strike the reader as being terribly inefficient of the herbivores to leave 90% of their dinner still on the table, so to speak. Why wasn't more consumed? Part of the answer is that evolution does not always favor efficiency. From the point of view of the plants, the less energy that gets transferred to the herbivores the better. It is worth noting that much of the plants' energy gets bound up in the form of cellulose—a very stable compound, and one which many animals cannot digest. Furthermore, it should be obvious that if *all* the vegetation were consumed, and *none* given over to reproduction by the plants, the animals would eliminate their food supply and become extinct. Somehow, a balance must be struck, but ecological theory is still unable to provide a satisfying account of how this occurs. Be this as it may, the 10% slice utilized by herbivores does not mean that the remaining 90% of plant mass is "wasted." Most of it will be consumed by decomposers during early winter, when cool, wet weather and winds will beat the dead herbage to the ground. Our clipping was at the end of the dry period.

Whatever the details of consumers, decomposers, and producers may be, the three appear to coexist in a state of balance with the microenvironment. We have seen, in previous sections, that as one moves inland from the lip, great changes in the grassland ecosystem occur. For one thing, many of the plant species change; for another, so does the microenvironment. But the magnitude of change in plant productivity was relatively minor over that distance.

How efficient is net plant productivity at Bodega Head? Efficiency in this case is equal to productivity divided by incoming radiation, both measured in calories (see Appendix B). Using radiation data for the nine-month period from the pyranometer.

$$\% \text{ efficiency} = \frac{2{,}493 \text{ kcal/m}^2}{958{,}035 \text{ kcal/m}^2} \times 100 = 0.26\%$$

An efficiency of 0.26% is quite low, compared with other grasslands. Although a pasture in the San Joaquin Valley, according to a study by Williams (1966), showed an efficiency of 0.09%, a perennial grassland in Michigan, according to a second study by Golley, showed one of nearly 1%. Efficiency in other vegetation types ranges between 1% to 5% (see Phillipson 1966). Why is efficiency so low at Bodega?

First, we cannot attribute the low value to an error in our estimate of the herbivore consumption. Even if they utilized 25% of plant matter (rather than the 10% we allotted them), efficiency would only be raised to 0.30%. Second, efficiency was not low because of low light intensity. Incoming solar radiation was twice that received by the Michigan field mentioned above (in terms of cal/m²/time).

Low temperatures could be responsible for the low efficiency. The San Joaquin Valley pasture and the Michigan field both experience much higher temperatures during the growing season than does the Bodega Head grassland. And up to a limit, plant growth is stimulated by high temperature. Low moisture could also have been responsible for the low efficiency. The San Joaquin pasture may not have been irrigated (Williams didn't say in his paper) during the dry growing season, but the Michigan field was subject to typical Michigan summer rains. There is no rainfall in summer at Bodega. All this implies to us that, given summer moisture and/or increase in temperature, plant productivity at Bodega Head could be much higher. Growth chamber experiments can check this hypothesis.

From what we have seen in the grassland chapter, we now have enough results to begin making some predictions. For example, the grassland plant community is likely to change in the near future to a lupine-dominated scrub with many perennials and fewer annuals, and the animal community will change with it. Many other predictions might be made, some of which might be useful for conservation policy. Is the Head to become a state park, experiencing heavy traffic by horses, vehicles or hikers? Then the easily compacted soils and the aggressiveness of introduced footpath weeds may be important considerations in predicting whether a park is a good idea or not. Is the Head to become a biological preserve? Then its value—in terms of biological uniqueness— should be compared to other potential sites in order to predict the benefits of that plan; and the effect of activity around the Marine Laboratory on the natural community should be taken into account. Should its offshore waters be utilized as an outfall area for heated discharge from a power plant? A yes or no answer depends upon evaluation of the temperature tolerance limits of marine organisms throughout the food chain. Should the Head be used for truck farming? Factors such as the effect of pesticides on nearby natural areas, and the nutrient status of the sandy soils, would enter into the answer. Should light industry be invited to Bodega Bay? The effect of atmospheric effluents on the harbor microenvironment (especially if inversions trap the pollutants—take a look at many coastal towns in northern California and Oregon) must be considered in reaching a decision. The list could be extended indefinitely.

3

Rocky Intertidal Habitats

General Structure of the Habitat:
Influence of Substrate and Waves

We shall approach the rocky intertidal area in a manner
slightly different from that used thus far, for several reasons.
For one thing, the rocky shores differ so much from any terrestrial habitat that a whole new series of questions needs to
be asked. Secondly, the community here is somewhat more
uniformly distributed than are many terrestrial ones, partly
because the habitat is readily colonized from the sea. For this
reason, and also because work on the Head itself has only begun, we shall rely rather heavily on work done elsewhere.
Finally, we shall approach the subject more from a structural,
and less from a physiological, point of view, merely because
one of us happens to have a strong interest in "ecological
morphology."

Fig. 3.1   General aspect of the rocky intertidal.

ROCKS AND WAVES

The term "rocky intertidal" is virtually self-explanatory. It designates areas of firm substrate which are periodically submerged and exposed as the tide rises and falls. Owing to the action of waves, the region just above and just below the rocky intertidal is subjected to comparable effects and may reasonably be treated as more or less the same habitat. Thus at higher levels, spray and waves during storms extend the marine influence upward, often by many feet. And the mechanical action of waves may be felt some distance below the lowest tidal level. The intertidal region may be looked upon as an ecotone, in which the properties of the terrestrial and the marine environments are mixed; yet the hard surface combined with water movements renders it a unique habitat in its own right. At Bodega Head most of the rocky intertidal zone may be classified as "exposed outer coast." That is to say, the rock along the western side of the Head receives the full impact of the waves. A few areas, in which outlying rocks break the force of the waves, are termed "semi-protected outer coast." The effect of waves is somewhat difficult to quantify, but one can at least say that waves are quite large at Bodega. The vast extent of the Pacific Ocean, combined with the pre-vailing wind patterns, causes waves to build up to great size, much larger than those usually experienced on the Atlantic coast. Any animal or plant which inhabits such a place must be able to cope with some very powerful mechanical forces.

Uneven erosion has produced a characteristic pattern along much of the shore. An originally rounded cliff has been undercut near sea level, leaving an intertidal platform or shelf of variable width, with many shallow pools and deep fissures (Fig. 3.1). The rock itself provides a firm place where ani-mals and plants can cling or cement themselves. It is relatively hard, and, unlike the rocks in some other areas, does not sup-port much of a fauna of burrowing animals. Many animals and plants, however, do find a certain amount of protection and shelter in cracks, fissures and pools, and under large boul-ders. Of equal importance in providing shelters are the many cavities produced by the organisms themselves. A host of small animals make their homes between and within attached ("sessile") animals and plants.

Although the action of the waves renders the exposed situations uninhabitable to many creatures, it does not much decrease the quantity of organisms. Indeed, by bringing in a rich store of nutrients and foods, and by making life difficult for predators and competitors, waves may greatly benefit those which can put up with the hard conditions. This is obvious from the thick mat of organisms growing there. Nonetheless, a certain amount of adjustment is essential, and the importance of wave shock here is manifested in the structure and life style of both animals and plants.

One of the most striking organisms adapted to extreme

wave shock is the sea palm, *Postelsia palmaeformis* (Fig. 3.2).
This alga lives attached to the rock by a very strong holdfast;
on the least protected crests of rocks one may see groves of
them, each with a stout, flexible stalk surmounted by a group
of leaf-like processes. Although the plant stands erect, the rub-
bery stalk bends with the waves. The holdfast is large and
very firmly attached: plants found on the beach after storms
are still fused to barnacles, indicating that it was not the plant
itself that gave way.

The attached, or sessile, way of life is appropriate for
macroscopic plants, since these need not pursue their food.
Many intertidal animals have evolved a sessile or at least sed-
entary mode of existence, made possible by their use of the
food coming to them. The movement of water substitutes for
pursuit and provides a basis for a substantial "industry." Attached animals are often "filter feeders" which, as the term
suggests, have a straining arrangement allowing them to remove minute nutritive particles from the water. This component of the intertidal fauna is exceedingly conspicuous, and
includes representatives of many phyla. On exposed surfaces
the bivalved mollusk, *Mytilus californianus*, the common mussel, forms extensive beds. These animals can move around a
little, but they attach themselves by a bundle of proteinaceous
threads called a *byssus*. Equally characteristic of such places
are the barnacles, which are crustaceans in which the legs have

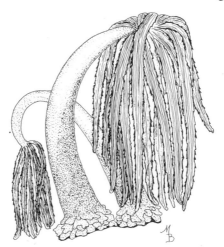

Fig. 3.2   Sea palm

been modified to form a food-gathering apparatus. The common "goose barnacle" at Bodega Head, *Pollicipes polymerus*, forms large, conspicuous clusters. It has a rather flexible stalk. The "acorn barnacles" (*Balanus* and some others) live entirely encased in hard plates and are firmly cemented to their susbtrate without an intervening stalk. On the whole they are fairly flat and conical, and thus present a streamlined profile which should help them resist being dislodged by predators and waves alike. In crowded situations, however, they tend to depart from this optimal form, becoming taller. Filter feeders of different groups tend to predominate in cracks and under rocks. Here one often finds a thick coating of sponges and tunicates.

On rocks and in tide pools one finds large numbers of sea anemones. One of these coelenterates, *Anthopleura elegantissima*, is very common on semi-protected rocks, such as in Horseshoe Cove. By asexual reproduction it forms clusters and even extensive mats. The larger *A. xanthogrammica* is abundant in tide pools. These anemones are primarily carnivores, whose stinging capsules allow them to catch small animals. In addition, *A. elegantissima*, in situations where there is sufficient light, contains symbiotic "zooxanthellae" and "zoochlorellae," unicellular plants thought to supply the anemone with a certain amount of food. Occasionally, for reasons not well understood, large numbers of fish are killed and wash up on the shore at the Head. At such times we have observed *A. elegantissima* with fish projecting from the mouth, suggesting that this species may be quite opportunistic in its manner of obtaining food.

The sessile plants and animals support a substantial number of herbivorous and carnivorous grazers. Yet the problems of dealing with wave shock impose strict limits upon how such animals can operate. Basically, they must be able to hang on and crawl around at the same time. Among the most abun-

Fig. 3.3 Tide pool showing sea anenome and associated organisms.

dant animals having the necessary features are certain gastropod mollusks (snails and slugs). The flat foot of these creatures suits them quite well to clinging, while their slow mode of crawling about presents no serious disadvantage in pursuing attached food items. In resisting wave shocks, however, the most effective arrangement differs somewhat from that of a conventional snail with its coiled shell and only a moderate-sized foot. Rather, the body needs a lower profile, like a dome, presenting minimal surface area; and the foot should form a large suction-cup. Such an arrangement has evolved independently in several groups of gastropods. When the shell is retained, but modified to become cap-shaped, the snail is termed a "limpet"—but again this form has evolved repeatedly.

One of the most successful limpet groups is the family Acmaeidae, the Bodega Head forms of which have all been

placed in the genus *Acmaea* (Fig. 3.4). We will retain this
conservative terminology, but it seems worth mentioning that
some taxonomists advocate splitting the genus into smaller
units. About a dozen species of *Acmaea* are known from Bodega Head. Although the ancestral stock of the genus would
seem to have been grazers upon the soft algae that live on the
intertidal rocks, the group has diversified extensively, especially
on the Pacific coast of North America. A number of species
have become quite specialized in where they live and what they
eat, and may be found only on a particular kind of animal or
plant. We will discuss them further when we deal with zonation later in this chapter. They are quite abundant on exposed
outer coast, and by clamping down they can resist predators,
wave shock, and the drying action of the air.

Another group of limpet-like forms is characterized by one
or more holes in the shell which provide for the egress of water:

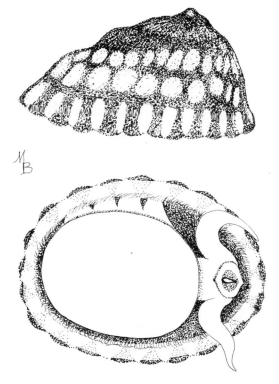

Fig. 3.4   Limpet. Side and bottom view.

for obvious reasons such an arrangement suits them poorly for resistance to drying. Among these the best known to laymen is the herbivorous red abalone, *Haliotis rufescens*, a popular food item now as it was with the Indians (see Chapter 7). A number of shell-less gastropods (especially "dorid" nudibranchs) have much the same habits. Dorid nudibranchs, which feed upon sponges, are oval, streamlined and flattened, with a retractile tuft of gills at the posterior end of the body (Figs. 3.5a and 3.5b). Many are brightly colored and can easily be mistaken for the sponge itself. A close look at certain red sponges reveals a small red nudibranch, *Rostanga pulchra*, which evidently obtains food protection and a place to lay eggs from its host. Another class of mollusks, the Loricata or chitons, have much the same outward form. Chitons are usually recognizable by their row of eight shells, surrounded by a tough yet flexible rim of mantle tissue. They are streamlined, and, unlike limpets, can bend their bodies to adjust their shape to an irregular surface. The more common chitons are herbivorous.

Another group of organisms which can both hang on and crawl about is the echinoderms ("starfish" and allies). Their system of tube-feet allows a slow, yet secure progression. The group includes some of the most important predators in the intertidal zone, especially at lower levels: they probably can-

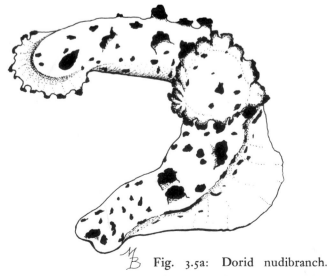

Fig. 3.5a: Dorid   nudibranch.

not withstand the desiccation higher up. The large, many-armed and highly flexible surf star, *Pycnopodia helianthoides*, moves quite rapidly for a sea star. It evokes rapid flight of many snails when it invades a tide pool. The five-armed, stiffer *Pisaster ochraceus* is found in considerable numbers at and below the lower limits of mussel beds: it eats lots of mussels and barnacles. A stouter sea star, the bat star or *Patiria miniata*, seems to be an omnivore. The common purple sea urchin, *Strongylocentrotus purpuratus*, is one of the few intertidal animals that eat larger algae. Extensive areas at the Head consist of beds of these animals, many of which live in small depressions which they have gouged out of the rock.

Vertebrates, because of their morphology, are predisposed to a motile existence and as such cannot tolerate the hammering and jostling of wave action and tend to move out or seek shelter. Many species of fish migrate into the intertidal zone when the tide is in. A few are highly adapted to intertidal life, such as the flattish, highly streamlined clingfishes with their strong ventral suction-cup, and have specialized in intertidal existence. In tide pools, even at higher levels, one can find small fishes of the family Cottidae, such as *Clinocottus*. So nearly do these mottled creatures match their background, that when sitting motionless on the bottom they virtually always pass unnoticed. Various eel-like fishes ("blennies") also reside in the intertidal; they are common under rocks. There is such an abundance of exposed food during low tide that, as

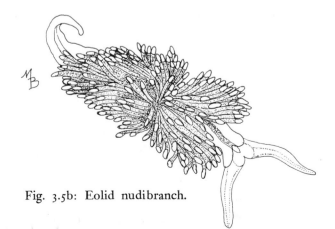

Fig. 3.5b: Eolid nudibranch.

Fig. 3.6 Pelagic cormorants

nature would have it, several species are adapted to exploit it. Birds which have the mobility to dodge the surf invade the intertidal. The black oyster-catcher, *Haematopus bachmani*, feeds entirely on invertebrates of the intertidal shelf. It feeds just above the water line, dislodging mussels, limpets and barnacles with its long, red, rectangular bill. Other rocky intertidal birds are willets, wandering tattlers, black turnstones, spotted sandpipers, and surf birds. These birds, all in the sandpiper family, feed on isopods and polychaete worms—animals which are easily removed from among mussels, barnacles and anemones during low tide.

A number of birds nest on rocks in or above the splash zone. The pelagic cormorant (*Phalacrocorax pelagicus*, Fig. 3.6) is the most conspicuous, with dark plumage and a long neck. Other species of cormorants frequent the rocks and the harbor as well. Cormorants are excellent fishermen, as are their close relatives, the brown pelicans, which are occasionally seen on the rocks but more commonly in the harbor. Captive cormorants are used by some Japanese fishermen to catch fish. The fisherman puts a ring around the cormorant's neck so it can't swallow fish, and as a point of attachment for a rope to haul him back after a successful dive on a fish.

Another common resident of the rocky intertidal is the pigeon guillemot (*Cepphus columba*). A member of the auk family, it squats like a penguin and displays bright, red feet. These birds generally feed offshore within a mile or two of the coast. They swim underwater using their wings as oars. Only members of the auk family and the penguins swim in this manner.

Western gulls are the scavengers of the area. They are abundant and eat a variety of animal matter and debris, either catching it alive or feeding on its dead remains. They also steal fish from pelicans and cormorants.

No reptiles and few mammals feed in this luxuriant habitat. A notable exception is the raccoon (Chapter 2) which feeds extensively on small crabs, as evidenced by the large quantities of exoskeletons found in their scats.

SOME SUBTLER EFFECTS OF WATER

By now it should be clear that wave action exerts a profound influence upon the lives of intertidal animals and plants, not only directly, through a need to avoid being swept away, but indirectly, through its effects on their food supply and predators. A number of additional ways in which water movement affects intertidal life deserve at least passing mention, particularly since many of these are overlooked by the casual observer. One often gets the impression that the main reason for mortality of intertidal organisms, apart from predation and the like, is the physiological stress imposed by exposure to the sun and air. Yet the surface water itself may bring in materials which impose hardships upon the intertidal organisms. Dayton (1971), for example, has stressed the importance of

intertidal logs. Especially during stormy weather, large pieces
of wood may be hurled like battering rams against the unpro-
tected animals and plants. The wood of course is concentrated
at the surface, and avoiding it may be one of the reasons why
some animals migrate downward in the winter. It is proble-
matical whether such mortality should be attributed to "pollution." A lot of the wood that washes up on the rocks at
Bodega Head appears to have been released by logging operations. On the other hand, a certain amount of driftwood is
used as fuel, and there were doubtless some logs in the water
before man intervened. Another form of mortality occasionally seen at the Head results from the stranding of open water
animals. Especially in the spring, the wind from the sea may
bring in large numbers of a small coelenterate, the "by the
wind sailor" *Velella*. This animal floats at the surface, carried
along by a low sail-like structure. From time to time large
numbers of them are washed into higher tide pools, where
they rot and the products of decomposition kill the inhabitants. Similar effects result when large numbers of fish die. In
the spring and summer, the water often becomes heavily populated by a microscopic dinoflagellate which is then eaten by
mussels and other filter feeders. The dinoflagellate contains
one of the most potent nerve-poisons known but it doesn't
harm the mussel, which is immune. On the other hand, the
poison accumulates in the mussels and people who eat them
when the level of poison is high may even be killed. The U.S.
Public Health Service maintains a quarantine when the mussels are dangerous, one which occasionally lasts into December. Virtually no work has been done on how such poisons
influence animals other than man.

TERRESTRIAL INFLUENCES

At increasingly higher elevations the conditions of life
become more and more like those on land, but they continue

to be of mixed nature. Hence, the organisms must one way or another be able to live in two sorts of environments.

Most of the larger marine animals are provided with gills. These structures generally do not work very well, if at all, in air, and various accommodations are necessitated by the fall of the tide. Some animals migrate or take refuge in pools. Others shut down and become temporarily dormant. Many can switch to aerial gas-exchange, and this maneuver is rather common in the intertidal. A number of animals, indeed, have switched to breathing air. A little lung-bearing slug, *Onchidella borealis*, is very commonly seen at low tide browsing on diatoms. The "periwinkle" of higher levels, *Littorina planaxis*, will leave if placed under water. Difficulties of respiration are further compounded by loss of water through evaporation. Many of the algae at higher levels have a branching, fuzzy aspect which is thought to aid the retention of surface-film of sea water. Comparable adaptations have been documented for animals. One of these is a marked tendency to gregariousness: water caught between groups of limpets, barnacles and sea anemones helps keep them all wet, and several together are more effective than a solitary individual in raising the local relative humidity through evaporation.

The bare rock of the intertidal may become quite hot when exposed to the direct rays of the sun. Changes in temperature can be very rapid, and the high temperatures accentuate any tendency to drying. Actual periods of temperature stress, however, are very hard to observe, as was found by Thomas Wolcott in a field study of *Acmaea* at Bodega Head. Mortality is greatest in the smallest animals, so that a large individual will ordinarily have already experienced greater stress than one is likely to observe. Mass mortality does occur, but mainly at the higher levels, and one needs to wait for a low tide on a windless, sunny, summer day. Such conditions are but rarely encountered at Bodega Head.

As if heat, high salinity and desiccation were not enough, intertidal organisms must also deal with an opposite sort of influence. The air in winter may drop below the freezing point, so that the minimum temperatures drop below those in the sea. In addition, rain may lower the salinity, and many organisms have little capacity to withstand the stress.

CONDITIONS IN POOLS: A CASE STUDY

Thus far we have simplified our discussion by focusing mainly upon situations in which the organisms are uncovered by the receding tide. In pools, the picture differs somewhat. Exposure to the drying action of the air is no problem in a tide pool, and gills remain in their proper medium. However, a tide pool should be viewed as more than a displaced chunk of the open sea. Rather, it constitutes a semi-enclosed body of water in which the moderating effects of circulation are reduced. Depending upon size and position, environmental influences may fluctuate considerably. The sun may raise the temperature of the water, and evaporation or rainfall may radically alter the salinity. In addition, the animals and plants may deplete the supply of oxygen or otherwise alter the chemical environment. The frequency and rate of exchange of materials may differ radically in high- and low-level pools.

The particulars of life in a pool are clear from a simple twenty-four-hour tide pool watch conducted by students at the Bodega Marine Laboratory beginning at 8 o'clock in the morning on 25 June 1969. The pool studied was located on the south side of Horseshoe Cove, at approximately 3.5 feet above mean lower low tide level. It was of moderate size and depth: about 3.9 m by 2.6 m at the surface and around .5 m in maximum depth. Several of the more common macroscopic plants were growing in it, indicating that conditions were reasonably moderate: *Phyllospadix, Pelvetiopsis, Cladophora* and coralline algae. The animals were the sort that are hardy but not found in the highest pools: sea anemones, sea stars and several species of snails, crabs and fish. Every hour the following were measured: air temperature, pool temperature both at the surface and at the bottom, and oxygen content of the water as determined by the standard Winkler technique. The action of waves and the general weather conditions were likewise monitored. The temperature data and tide minima and maxima are graphed in Fig. 3.7. It was a clear day, with only occasional cloud cover. In the morning the tide was low and wind was slight, so waves did not enter the pool until

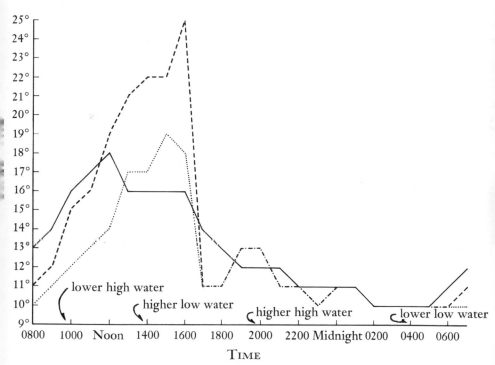

Fig. 3.7 Air, surface water and bottom water temperature corellated with tide level (temperature in degrees Celsius). Unbroken lines show air temperature; dashed lines show water temperatures at surface of pool; dotted lines show water temperatures at bottom of pool.

around 5 P.M. Air temperature in the early morning was 13 °C and rose to a cool 18 °C around noon, declining to a nighttime low of 10 °C. Even in the deepest part of the pool, water temperatures peaked at 19 °C, which is much warmer than anything occurring offshore. Because warm water tends to rise, the surface layer was much warmer, beginning at 11 °C and rising to an impressive 25 °C; it was evidently still climbing when waves began to come in and temperatures dropped precipitously to around 11 °C. During this period the plants were photosynthesizing, and oxygen levels climbed from about 7 ml/l to 17 ml/l. They returned to around 7 ml/l when waves entered the pool. Thereafter, temperatures remained fairly constant, not much above those of the nearby sea. After 7 P.M., i.e., in the dark, oxygen levels declined still more, reaching a low level of about 2 ml/l from 2 A.M. to 5 A.M. The figures given above are more or less what one might expect, but it should be borne in mind that each pool and each day is unique.

Intertidal Zonation

As the intertidal region contains a steep gradient between two strikingly different habitats, one would only expect different kinds of organisms to replace one another with increasing elevation. Yet even more impressive is the phenomenon of "zonation" in the stricter sense: whole assemblages of plants and animals occupy narrow bands, each with a distinct level and with definite breaks in faunal and floral composition where they meet. This is true even though wave action may displace the zones upward to some extent. How many such zones one should recognize and where to draw the lines is subject to dispute, and different arrangements may be necessary to cover the variety of localities. Nonetheless, authorities seem to agree that the phenomenon is "real" in the sense that the groups are not just arbitrary or subjective. A great deal of descriptive work, and many classification schemes, are available (Colman 1933; Doty 1946; Hewatt 1937; Hedgpeth 1957; Lewis 1964; Ricketts, Calvin, and Hedgpeth 1968; Stephenson and Stephenson 1949).

Exactly what causes organisms to display the observed patterns of zones remains an open question. From the beginning it has been clear that physical factors are somewhat important. Regions above and below the part of the shore covered and uncovered by the tides clearly have conditions of existence different from those between, but this does not of itself reveal the mechanisms which determine who lives where. Traditionally, it has been tacitly assumed that an ability to survive heat and drying is most important. Yet it is ever more apparent that other influences, especially biological ones, must be taken into account. The longer an aquatic animal is under water, the more time it can devote to feeding, and, at the same time, the greater the threat from aquatic predators. New ideas in ecology are now revolutionizing the traditional picture. Until very recently, ecologists have underemphasized competition, some even denying that it exists. As we mentioned in the Introduction, it now seems clear that many animals and plants are excluded by competition from places where they could otherwise survive or even flourish. Another valuable insight comes from theorizing about environmental stability. One might hypothesize that mass mortality due to such influences as the weather would occur more frequently at one level than another, probably more often at the higher levels. If so, this might reduce spatial competition, favor species that can colonize the depopulated regions most effectively, and perhaps reduce the diversity of organisms. This would follow, at least, from currently popular notions; but we should be aware that much of this theory remains unsubstantiated by empirical work. In the following discussion we hope to cast some light on the degree to which zonational patterns accord with both the older and the newer ideas. Additionally we will use some of the biological peculiarities of the inhabitants from different levels to suggest that more could be added to the total picture. The different zones seem to be characterized by different kinds of natural economies, so that the inhabitants make their livings in different ways. This can best be seen in the limpets, which will be stressed here, but other groups display comparable phenomena. We will attempt to enlarge upon such themes in the section on mudflats.

To understand why the breaks in faunal and floral composition occur where they do, it helps to consider the topic of "critical levels," and this requires that we discuss the tides. It is common knowledge that tides result from the gravitational attraction and the relative movement of the earth, sun and moon. The moon, and to a lesser extent, the more massive but distant sun, cause the water to rise on the opposite sides of the globe, such that there will be two zones of high tides and two of low tides, with the highs opposite to highs and the lows opposite to lows, at any one time. As a result, the rotation of the earth brings about slightly less than two high tides and two low tides a day—the "diurnal cycle." Conditions of the local terrain vary, however, so that in some places there is but a single high and a single low tide, or even virtually no tide at all. The Bodega cycle, with a "higher high" tide, a somewhat lower "lower high" and comparable "higher low" and "lower low" tide each day, is quite common, and it is the usual pattern along the West Coast of North America (see Fig. 5.3). A critical level is a point at which the amount of submergence and exposure changes abruptly. One such critical level should be self-evident: above a certain point, organisms are covered only by "higher high" tides and by wave action. Both the length of time exposed and the immersion period differ considerably from those at lower levels.

To the diurnal cycle, we must add the consequence of the twenty-eight-day lunar cycle. The moon revolves around the earth, so that sometimes its gravitational attraction pulls with the sun's, sometimes against it. When the three celestial bodies are in line (twice per cycle, or every other week), low tides are lower, high tides are higher. Tides at this time are referred to as "spring" tides, a term derived not from the season of the year, but rather from the same root: to jump or well up. When the moon pulls at right angles to the sun, we get less extreme rise and fall, and tides at this period are called "neap tides." Again it should be apparent that the highest areas will be covered, and the lowest exposed, only occasionally. To all this we must finally add that the annual cycle has some effect. Extremely low and high tides occur in May and June, and in November and December; less amplitude characterizes March and April and September and October. Tides come later each day by about fifty minutes. And the time of day at which lows and highs occur may have considerable biological importance. Low tides in summer at Bodega Head would be much harder on intertidal organisms were it not for the fact that they tend to come in the early morning.

Experts who have tried to set up classification schemes for intertidal zones seem uncertain as to whether the means or the extremes are the more important in drawing boundaries. In his classical paper on the vertical distribution of algae, Doty (1946) found most of the breaks at the extremes—such as the lowest lower low water, i.e., the lowest point ever uncovered. Above this level, obviously, organisms unable to withstand exposure at all are excluded. Similarly, above the critical level of the lowest higher high water, the organisms may have to endure days on end of continuous exposure.

Analogous considerations apply to a number of other levels, but one exception has always constituted a puzzle: the very conspicuous discontinuity at "mean lower low water."

One possible solution has lately been worked out by one of us (M. T. Ghiselin) in collaboration with David A. Cobb. It all hinges on the fact that the mean in this case is also a "mode" —the tide fluctuates about this level. On an "average day" the area below is not exposed; above, it is exposed. Comparable considerations would apply to other "mean" levels, especially at mean higher high tide. If one simply counts the total number of hours of exposure over a year, or if one assesses the length of time exposed, no breaks occur at the mean. Nonetheless, the conditions on an "average day" are important, because they reflect the conditions usually experienced by an organism. For feeding, what matters is the regular, day-to-day availability of food, not just how much time is available for gathering it up. To use an analogy, people do best on two or three square meals a day: the same number of calories presented in the form of one meal a week alternating with a six-day fast wouldn't work too well. Hence, one might expect the organisms living below a mean level to be adapted to usually being submerged, those above to usually being exposed. We shall suggest ways in which this idea might be applied to concrete situations later on. It bears repeating that older views on distribution tended to overemphasize physiological stress and to underemphasize feeding and competition. As we shall see, different feeding mechanisms seem to predominate at different levels; and sometimes diversity changes while at others one species is simply replaced by another member of its genus.

The most widely used scheme for classification of intertidal zones on the West Coast of North America is that of Ricketts and Calvin (Ricketts, Calvin, and Hedgpeth 1958, and earlier editions). It compromises between various other schemes, and uses a combination of means and extremes. They number the zones 1 to 4, from above to below. Their uppermost zones (1) consists of that portion of the shore which is wetted by waves and spray, and, in the lower portion, by the higher high tides. In other words, the upper limit is somewhat indefinite and the lower is delimited by the point reached by the highest lower high tides. Not only is this area covered no more than once a day, but during neap tides it may not be covered at all. Zone 2 extends downward from this point to the mean higher low water level, zone 3 to the universally-accepted mean lower low water. Finally, zone 4 extends to the extreme low water level, where a "subtidal" zone begins. In the following discussion we will follow Ricketts' and Calvin's system.

ZONE I

The uppermost part of this zone is wetted only by waves and spray, while, lower down, the area is periodically inundated by the highest tides. Wave action of course varies with the weather. During storms even quite high areas receive at least an occasional drenching. A number of fresh-water seeps at higher levels support a green alga, *Enteromorpha*, and crustose lichens are abundant on the rocks. But most of the plant life consists of minute algae which give the rock somewhat of a greenish or blackish cast.

Zone 1 animals are numerically not every impressive; they represent a few species, and although they may compete for the best positions, signs of crowding only begin at the lowest part. At the higher elevations one may observe an iso-pod crustacean, the rock-slater, *Lygia occidentalis*, rapidly running about. Like the terrestrial "pill bugs" to which it is related, it loses water fairly easily and seeks refuge in moist situations under boards and stones. The most abundant macro-scopic animals in this zone are herbivorous gastropods. A peri-winkle, *Littorina planaxis*, lives only in this zone; at lower levels its range overlaps with that of the smaller *L. scutulata* which is better considered a zone 2 form. The several species of limpets which inhabit zone 1 would all appear to have much the same diet, for there is little food available to them but the microscopic algae on the rock and on other animals. But they seem to have subdivided their niche, insofar as they have different ways of resisting desiccation. *Acmaea persona* can be found on the fully shaded undersides of boulders, from whence it emerges to feed under optimal conditions. *A. digitalis*, on the other hand, occupies a more exposed position on open rock faces, although it congregates in holes and cracks. According to Frank (1965), this animal lives at a higher level in the winter, and the larger individuals tend to occupy a higher position. They can fit fairly closely to the rock surface. When placed under water in an aquarium, they crawl out of the water where they remain even until they die of desiccation. In the normal habitat, wave action would save it. At the lower part of zone 1, another limpet, *Acmaea scabra*, becomes common. This one is very flat, and may form a scar on the rock or take advantage of a natural depression into which it fits quite tightly, at once resisting desiccation and becoming hard to dislodge. This animal returns to its home after foraging nearby when the tide is in. *A. scabra* also occurs in zone 2, and its uppermost limit coincides with that of the common barnacle *Balanus glandula*. It may be no coincidence that the limpet bears a striking resemblance to the barnacle, for this would be an effective form of camouflage. John Sutherland devoted several years to a comparison of *A. scabra* in zones 1 and 2 at Bodega Head (Sutherland 1970). As one might expect, the two groups had substantially different features, largely attributable to there being less food but more space and more frequent catastrophic mortality above. At higher levels, the animals were fewer but larger, and they grew faster. Recruitment was slower and reproduction was concentrated in the winter when food was more abundant. In zone 2 recruitment rates were higher, and the animals, although more numerous, grew more slowly. By reducing the population density, and thereby making more food available to the remaining individuals, Sutherland was able to show that lack of food was keeping the biomass down. Predation turned out to be more intense at lower levels, and reproduction was less seasonal.

Sutherland's work neatly complements earlier studies by Castenholz (1961), who noted higher algal productivity during the winter at upper levels owing to reduced desiccation. After completely eliminating the herbivorous gastropods in the lower levels, Castenholz found that the minute plants became much more numerous. Such experimental work clearly demonstrates that at least some marine herbivores are limited by their food supply; that food is limiting for terrestrial herbivores has been denied by Hairston, Smith, and Slobodkin (1960), but some ecologists view their theory with profound skepticism.

The tide pools of highest levels present some interesting peculiarities. They are often very fertile, because bird feces are carried into them, but salinities vary enormously (indeed, such pools may be saturated with salt, and often dry out completely). They are often colored green from dense swarms of a microscopic alga, *Dunalliella*, and they swarm with a minute, reddish copepod crustacean, *Tigriopus californicus*.

ZONE 2

At this level, both the number of individuals and the number of species increase markedly over zone 1. Rather than lose the reader in a mass of details, we shall discuss a smaller part of this total diversity for this and subsequent zones. The greater numbers, one would expect, should create more crowded conditions and, therefore, intensify competition for space. And longer periods of submergence ought to favor the existence of filter feeders and the larger algae. The former feeding type is largely represented by the common barnacle already mentioned, *Balanus glandula*. Snails which feed upon it, such as *Acanthina* and *Thais*, also live here. Other barnacles and mussels begin to appear at this level, but these are best considered zone 3 organisms. Doty (1946) found several zones of macroscopic algae within zone 2. Many present a rather "bushy" aspect and are not exceedingly large, and both features may be important in resisting desiccation.

High in zone 2, and somewhat extending into zone 1, one finds a community, the "*Endocladia muricata–Balanus glandula* association," which was studied in great detail by Glynn (1965) at Hopkins Marine Station in Pacific Grove. The association is named after the plant and the animal which may be considered the "dominant" species in two senses of that term. In the first place, they are dominants in the *quantitative* sense; they form much of the bulk of living material. Secondly, they are dominant in a *qualitative* sense, in that their presence determines the structure of the community. Quantification for many is such a shibboleth that the importance of it in understanding what is going on may perhaps be overlooked. The spaces within the tufts of algae, and between the living barnacles and within the dead ones, shelter a large number of small organisms, both resident and transient. When Glynn began his study, he accepted the usual dogma that so exposed a community would be very simple, but it soon became obvious that a host of animals were simply inconspicuous. He found ninety-three benthic species, of which thirty-five were common within the community. Several species were previously unrecorded from the Eastern Pacific, and five species turned out to be unknown to science. One of these species, a tiny slug, represented a family previously known only from the Western Pacific, Indian and Atlantic Oceans. Several minute bivalves are found here, including the young of the common mussel; and a number of small barnacles, gastropods, crustaceans and even insects are common.

The association serves as a kind of nursery for juveniles of the lined shore-crab, *Pachygrapsus crassipes;* adults of this species are conspicuous on the open rocks and in fissures. They are the more common crab in and around tide pools of zone 2, but range into zones 1 and 3. In tide pools and in rather moister situations one finds anemones (*Anthopleura*), a hermit crab, *Pagurus hirsutiusculus,* the herbivorous snail *Tegula funebralis* or topshell, and a number of forms common in the open lower down. We have already mentioned two limpets which extend into this zone from above: *Acmaea scabra* and *A. digitalis.* In zone 2, we see the beginnings of a new way of subdividing the limpet niche. *A. paradigitalis* and *A. scutum* feed on the microscopic algae, much as we saw above. But *A. pelta* bites off chunks of the larger macroscopic algae. *A. asmi* lives on the shells of *Tegula,* a snail.

ZONE 3

This, the second of our two middle zones, extends downward to mean lower low water. Within this zone, Doty (1946) found a distinct break in the macroscopic algal flora at the level of the lowest higher low water. The algae forming the higher-level subgroup, albeit taxonomically different, are about as diverse as those of zone 2; and their aspect is much the same. Below this point, there are many more species, and they tend to have a different physiognomy. We begin to pick up large kelps such as *Egregia*, and also coralline algae—forms with a large amount of mineral substance which evidently helps protect them from herbivores.

By contrast, animal diversity picks up considerably right at the boundary between zones 2 and 3. Many from above continue their ranges downward: *Pachygrapsus*, hermit crabs, and some of the limpets. The herbivorous "topshell" *Tegula funebralis,* on the other hand, is replaced by another member of its genus, *T. brunnea,* and along with it a different "limpet" hitches a ride. This "slipper" limpet, *Crepidula adunca*, is a filter feeder, unlike the *Acmaea* which scrape the surface, and may be treated as representative of the sudden increase in filter feeders within the zone. Unpublished work by Steven Hoffman shows that *C. adunca* does rarely occur on *T. funebralis*. A larger number of slipper limpets may be found on dead shells of *T. funebralis* occupied by hermit crabs. This study provides additional evidence that submergence is the important influence, for the crabs spend more time under water than do the living topshells. A new assemblage of herbivorous mollusks may in part depend upon the composition of the flora. Several species of chitons are common grazers here. One of these, *Tonicella lineata*, eats coralline algae. The zone 2 *Acmaea* species penetrate down this far, and we also get a couple of trophic specialists: *A. mitra* lives on coralline algae, while *A. insessa* is found exclusively on the stipes of *Egregia*, a large seaweed.

The most obvious zone 3 community at the Head is one called the *Mytilus-Pollicipes Pisaster* association, or what we shall loosely speak of as mussel beds. These have a very definite structure, largely determined by what they eat and who preys upon them. The mussels (*Mytilus californianus*) and the barnacles (*Pollicipes polymerus*) grow close together, and their gregariousness may provide some protection. Dayton (1971), as already mentioned, maintains that floating logs tear off portions of such assemblages. If one examines a mussel bed that occupies an extensive portion of a steep slope, it is clear at a glance that high up the mussels are much smaller, and that really big ones are found only below. We won't go into the details of why this is so, for it involves settling rates, ability to withstand desiccation, the availability of food, and the action of predators. The bigger mussels seem to be somewhat better than the smaller ones at withstanding the attacks of boring snails (such as whelks of the genus *Thais*) and of starfish

(*Pisaster*). Recent work increasingly implicates predators as the influence which determines the lower limits of mussel beds. Our own naturalistic observations on certain assemblages of *Mytilus* inhabiting lower levels on the intertidal shelf at Bodega Head suggest some interesting additional possibilities. These small beds occur at a level where *Pisaster* preys actively on a variety of animals, including mussels. Here the *Mytilus* are quite large for the most part, and they live in modest-sized clusters perhaps a meter or less across. Some of these clusters consist almost exclusively of large mussels. Others are made up of large individuals on the side *away* from the waves, while *facing* the open sea one finds a band of *Pollicipes*. Small *Mytilus* live among, and just behind, the barnacles. The natural inference would be that the larger individuals of *Mytilus* can fairly well withstand the attack of predators, since at the rear of these assemblages they are found right next to starfish. The *Pollicipes*, some of which live as pure stands, evidently provide something of a barrier to the predators, even though, as we have observed, starfish do prey upon them. Both Dayton (1971) and Paine (1969) agree that *Mytilus* will crowd out *Policipes* when the starfish are excluded. The barnacles may, however, also enjoy an ability to handle other factors such as wave shock more effectively than do mussels. We might further propose that a regular succession obtains, with the barnacles creating conditions which allow the mussels to take over. The *Pollicipes* provide a home for a beautifully camouflaged limpet which Giesel (1970) considers a genetic variant or morph of *Acmaea digitalis*— comparison of their radulae ("teeth") leads us to suspect that it may be another species. Be this as it may, we have here one more example of the genus becoming ever more specialized with respect to feeding and substrate at increasingly lower levels.

In the spaces between the mussels one can find a rich assemblage of small animals that make their home here. There are in great abundance various small crabs and other crustaceans, and a number of worms representing the phyla Platyhelminthes (flatworms), Rhynchocoela (an important group, but one so little known to the layman that it has no common name) and Annelida (segmented worms, like the common earthworm). The flexible, elongate worms are particularly well adapted for crawling about among the spaces and are mainly predators. Some of the crustaceans are filter feeders. A structurally rather degenerate pea-crab, *Fabia*, occurs inside the gill cavity of many mussels, and eats the food which the mussel gathers on its gills.

ZONE 4

This zone extends from mean lower low tide to extreme lower low tide level; beyond is the "sublittoral zone." In zone 4 exposure to air is both infrequent and brief. The biota contains three major elements: intertidal forms occurring above as well as in zone 4 itself; subtidal forms capable of withstanding brief exposure or avoiding it by migration; and, finally, organisms largely restricted to this zone. We shall now examine each of these groups in turn.

As to the first category, the remarkable point is how few species actually do range downward from the upper zones. Doty (1946), again, noted a complete replacement of macroscopic algae, although occasional individuals might be expected to cross the line. The animals display much the same pattern. McLean (1962) in a pioneering study of *subtidal* communities south of Carmel, California, found many species of common mid-level (zones 2 and 3) animals occurring at very low densities, for example the free-ranging *Acmaea pelta* (Fig. 3.4) and juvenile *Mytilus californianus*. Many forms occurring in zone 3 are actually displaced upward from zone 4, and others are even basically subtidal creatures. It should be obvious that organisms which prosper while submerged, yet

which live at higher levels, are at least physiologically capable
of existing lower down; and the fact that many do occur subtidally demonstrates that a "need" for exposure does not set
the lower limit to their ranges. Hence, one must seek another
explanation. It is becoming increasingly clear that the ability
to handle the conditions at higher levels, especially physiological stress and the brevity of time available for feeding, is
purchased at the price of a lessened capacity for dealing with
predation and with competition from the inhabitants (Connell 1970; Dayton 1971; Paine 1969). Both *Mytilus californianus* and *Tegula funebralis*, for example, seem to be decimated
at lower levels by starfish.

Our second group, forms that are better considered subtidal rather than intertidal, would form a very long list. Sometimes one species merely replaces another of the same genus,
but the difference can involve taxa of higher rank, even phyla.
The midtidal *Pagurus* species are replaced by *P. hemphillii*—
which abounds subtidally. The shore-crabs *Pachygrapsus crassipes* and *Hemigrapsus nudus* give way to crabs of the genus
Cancer—again, common subtidally. Although such echinoderms as *Pisaster ochraceus* range even into zone 3, whole
groups with but occasional exceptions begin to appear only
in zone 4, and these too extend downward: various species of
sea stars, brittle stars and sea cucumbers. A number of groups
of more or less soft-bodied or otherwise delicate sessile animals enter the picture in zone 4, where to some extent they
take over the broad niches occupied mainly by barnacles and
mussels above. The groups in question include many kinds of
coelenterates, sponges and tunicates. Along with these one begins to find the animals which feed upon them, such as the sea
slugs mentioned earlier in this chapter. Keyhole limpets, abalone, and some but not all chitons provide examples of grazing animals that extend upward from lower depths only part
way into the intertidal.

A number of plants and animals can be considered truly

zone 4 organisms. At the species level, such organisms are numerous. One example is an alga, *Laminaria sinclairii*, which is replaced subtidally by *L. andersonii*. On its stipe one may find yet another highly-specialized limpet which lives nowhere else, *Acmaea instabilis*. Many lower level algae give an impression of unusually large size, but the subtidal forms include the largest ones.

A common plant that provides a good "indicator" for zone 4 is the "surf grass" *Phyllospadix* (Fig. 3.8). It too harbors one of those highly-specialized limpets, *Acmaea paleacea*, a minute, elongate creature just wide enough to fit onto the narrow blade of its host. Although not a grass in the technical (taxonomic) sense, *Phyllospadix* does have grass-like leaves. It is a flowering plant, albeit highly modified since its ancestors reinvaded the sea from the land. A perennial plant, its stubby roots branch from rhizomes and anchor the plant firmly to the rocks. Surf grass dominates much of the upper half of zone 4; in places all the surface is occupied by surf grass, and the only other organisms present are algae and small animals on the leaves or among the rhizomes. The plants are so dense, and their rhizomes so intertwined, that it is difficult to judge

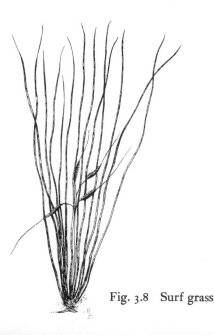

Fig. 3.8   Surf grass

the number of individuals per unit area. Possibly all the plants over a distance of several feet are vegetatively connected.

Very little is known about the ecology of marine flowering plants, partly because it is so hard to get at them. It is evident from even the most casual observation that the surf grass at Bodega Head grows only along the exposed outer coast, being absent from such calm areas as the harbor. One naturally wonders what it is that restricts this plant to areas where surf is intense. Some botanists think that mechanical agitation, produced by the waves, somehow is necessary to maintain plant vigor. But again, a correlation does not necessarily imply a cause-and-effect relationship. We suspect that the surf provides a cleansing action. We have maintained surf grass in aquaria for several weeks (see Appendix B) without mechanical agitation, but we did find it necessary to wash the plants once a week of any debris which settled onto the leaves. Vigor declined if the film was allowed to build up. But these experiments treat only one of the many possible influences.

We have tried to measure the growth of *Phyllospadix*, both in the field and in the laboratory. A number of low tides in the winter and spring of 1970 allowed us to revisit the same clumps of surf grass and apply the same clipping technique which was used for grassland (Chapter 2 and Appendix B). Along a portion of coastline, twelve two-foot-square plots were selected in the zone, and all surf grass leaves in them clipped back to a uniform stubble height of 5 cm. These initial clippings were collected, dried, and weighed; the weights indicate the "standing crop" of leaves.

Table 3.1 shows that the weight of standing crop varied considerably from plot to plot. But the variation did not correlate with the direction the plants were facing. Consider how the aspect of a hill (that is, which way it faces) affects light intensity and temperature at its surface. A south-facing slope is exposed to the sunlight for a longer time each day than a

Plot no.	Aspect, slope (%)	Leaf weight first clip (g)	Leaf weight second clip (g)	Regeneration (%)
1	west, 50	49.7	2.7	5.6
2	west, 50	65.5	3.6	5.5
3	southwest, 100	102.4		
4	southwest, 100	61.8		
5	west, 100	49.1		
6	west, 100	59.8	2.4	4.0
7	north, 100	53.9		
8	north, 100	56.3		
9	northwest, 50	67.2		
10	west, 0	133.3	5.0	3.8
11	west, 0	91.4		
12	southwest, 50	61.7		
average		71.0		4.7

Table 3.1   Original standing leaf crop (in grams) and regeneration in a twenty-eight-day period for surf grass in field plots in the intertidal. Aspect and slope of the plots are approximate only. Date of first clipping was 5 January, second clipping was 2 February, 1970.

north-facing slope, and it is more nearly at right angles with the sun's rays. Consequently, there is higher light intensity and higher temperature during the day on a south-facing slope. Plants should grow more rapidly as light and temperature increase, but as shown in Table 3.1, aspect has relatively little effect on the growth of surf grass. The average weight of the samples from plots 3, 4, and 12—all facing southwest—was 75 g, while the average weight from plots 7, 8, and 9—all facing north or northwest—was 59 g, only 22% less. The average standing crop was about 300 grams per square meter.

After twenty-eight days, the plots were revisited, and four of them were again clipped to a stubble height of 5 cm. These clippings were collected, dried, and weighed; the weights represent regeneration, or productivity, in the twenty-eight-day period. Table 3.1 shows that productivity, as percentage of the standing crop, averaged less than 5%. Again, differences did not correlate with the type of exposure. The 5% represents wintertime regeneration. We believe, however, that growth in summer, when water temperature is a few degrees warmer (see Table 1.1 and Table 1.2) would be faster. It seems reasonable that local differences in temperature at any one time are not particularly significant in determining growth rates, since the water temperature in exposed situations at this level should be approximately uniform.

To test the effect of temperature, light intensity and salinity on the growth of surf grass, we removed clumps of it from the rocks and grew them under controlled conditions in laboratory aquaria (see Appendix B). In general, we found wide tolerance limits to these factors. Growth did decline as temperature increased, for example, but it declined very little for so large a change in temperature. Water temperature of 70°F is not normally encountered by surf grass anywhere in its range; such a temperature occurs only far to the south of its distribution limit in Baja California. Yet, at this temperature, growth rate in the laboratory was only 25% less than at 55°F—the average Bodega water temperature. Something other than the effect of temperature must be restricting its range in the south.

Before closing our account of intertidal life we should reemphasize that we have treated but a very small sample, especially for the lower levels. In zone 4 and below, we begin to come across the remarkable diversity that characterizes the sea as a whole. This diversity is of a different kind than what we see on land, being a diversity at levels in the hierarchy of classification above that of the species. The land, in spite of the fact that it makes up only 20% or so of the earth's surface, is far richer in species than is the sea—thanks largely to the vast numbers of insects. But in terms of orders, classes and

phyla, the terrestrial fauna must be looked upon as impover-
ished. Only a few of the approximately thirty phyla have become really successful as free-living animals on land: arthropods, vertebrates and mollusks (represented by snails and
slugs). The sea, however, contains members of every phylum.
And many phyla are exclusively marine: echinoderms, Pogo-
nophora, Ctenophora and Chaetognatha, for example. Many
very important groups of marine organisms are not even mentioned in some college-level introductory biology courses.
Had we done full justice to the marine diversity, we would
have been defeated in our efforts to make this book intelligible
to a broad audience.

Conclusions

We have seen how the organisms replace each other progressively, and somewhat discontinuously, within the intertidal zones. Both biotic and physical factors are involved. A
full understanding of the total scheme of things, however,
will require much further study. Our limpets display a pro-
gressive trend, with high-level forms subdividing the niche in
terms of the manner of resisting desiccation, mid-level forms
eating food of different size, and those still lower down mani-
festing an extreme specialization in substrate. Similar phenomena could be adduced for other groups, all of which point to
differences in the natural economy between each level; and
indeed we will provide additional examples in other chapters.
It should be possible to devise experiments by which we could
tell how animals and plants are limited to their characteristic
situations—experiments like the exclusion experiments of Castenholz and others. Yet how much would they tell us? That
removing the limpets evokes a bloom of algae says something
about how one kind of organism limits another. But it provides little insight about why the diversity of limpets on the

Pacific coast is so great, or why the limpets belong to the genus *Acmaea* rather than some other group.

For the present we simply do not know to what extent we are working on materials which represent the conditions under which the contemporary fauna and flora evolved. Indians exploited the intertidal populations for millenia, but they had poor access to much of the habitat. More importantly, it is clear that the exclusion of the sea-otter has profoundly influenced community composition. In central California sea-otters have lately been recovering from their virtual extinction, and are now moving northward. They have fed upon their natural food, mainly such herbivorous invertebrates as sea urchins and abalone, eliciting in return a luxuriant growth of macroscopic algae. We have reason to think that their activity maintained kelp beds along the California coast for long periods. Just what do we learn from a study of disturbed ecosystems? Experimentally, or by accident, a predator or an herbivore is removed, and the system responds. But the system did not develop under such circumstances. The animals and plants are adapted to the conditions under which their ancestors long existed. Hence, showing how organisms relate to alien environments may provide us with a very warped conception of what is going on. Removing a predator, such as starfish or sea-otter, may profoundly affect the composition of a community, especially its species diversity. So what? To apply an analogy: were you to cut out somebody's heart, he would stop thinking. But if someone asks "Why do we think?," you don't reply "Because we have hearts." True answers are not always correct ones.

Even when studying perfectly natural situations, we need to keep proximate and ultimate factors distinct. It may be true, for example, that surf grass is limited to exposed situations through the necessity of being cleansed by the waves. But how much of an explanation is that? Does this really tell us why *Phyllospadix torreyi* lives where it does? Or does it merely give one "reason" why it cannot live elsewhere? Other marine "grasses" do quite well in quiet water. One way or another, organisms handle the environmental influences of their normal habitats. A plant which, for whatever reason, lives only in surf-swept places, might naturally take advantage of the cleansing action of waves. Yet such a phenomenon tells us nothing about why it doesn't live in quiet water, relying on some other mechanism of keeping clean—such as an animal which would use the deposited materials as food. These are evolutionary questions, ones which we are as yet unprepared to answer.

4

The Strand
and Dunes

North of Mussel Point, the rocky intertidal shelf abruptly gives way to a gentle sloping sandy shore. The break in topography marks the southern boundary of the San Andreas Fault (see Figs. 1.1 and 1.2). Great amounts of sand have been washed up beyond the high tide line and then blow inland to form the strand (beach) and dunes area, which stretches eastward all the way across the peninsula and northward all the way to Salmon Creek.

A low foredune, built up by constant deposits of sand and the congealing power of plant roots, separates the strand from the central dunes. Several hundred meters farther inland, the central dunes are bounded by a high hinddune. Behind its protection lies a hillocky region with a diversity of plant and animal life much greater than that of the stark central dunes and beach.

Fig. 4.1 Aspect of sandy beach.

Gradients in the Microenvironment

The strand and dunes cover about one-third of the area discussed in this book. This is about the same area as covered by grassland, but the differences between dunes and grassland are enormous from a biological point of view. The grassland exhibits a great variety of living things whose growth is so rank and numbers so dense that only 2% of the ground is not covered; but the dunes support very few species and as much as 80% of the ground is bare sand.

Soil texture is one reason for the biological differences. The dunes are composed of nearly pure sand (see Table 2.1); there are few clay particles to retain moisture or nutrient ions.

Fig. 4.2   Aspect of sand dunes

The sandy soil allows rain water to percolate rapidly through it, leaving the surface dry to a depth of one foot and unsuitable for seedling establishment. (Though sandy, the dune soil is not uniformly coarse throughout: in the central dunes, the fraction of soil which is coarse sand [diameter 1–2 mm] varies from 10% to 27%; behind the hinddune, coarse sand is only 1% or less.)

The terrain is low, allowing onshore winds with salt spray to sweep the area. The lack of plant cover permits wind erosion and sand blast, both of which impede plant and animal activity. The soil is low in nitrogen, there is no rich, black color from decomposed organic remains such as one sees in grassland soil. Indeed, the average nitrogen content of Bodega strand soil is only 0.0024%—even less than that of desert soils. Further, dune topsoil is strongly alkaline, averaging pH 8.4, in contrast to the slightly acidic grassland soil (Table 2.5).

The relative severity of environmental influence changes along gradients within the habitat. The conditions of life become increasingly severe toward the shore, and the distribution of plants and animals changes in a parallel fashion, just as in the grassland and in the intertidal zone. To sample these gradients, we established a 1,550-meter-long study transect extending from the strand, back through the central dunes, to the protected area behind the hinddune (see Fig. 1.1). Weather stations for collecting temperature data were established at three points along the transect: on the strand, just behind the foredune, and 1,000 m inland behind the hinddune. Salinity and plant cover were noted periodically along the entire transect.

TEMPERATURE AND WIND

When one walks the 50–100 m necessary to cross over the foredune, he feels a dramatic change in temperature and wind speed. On a typical day (even in summer) when walking on the beach, a jacket and a brisk stride are necessary to keep warm; but in the lee of the foredune, it often gets warm and calm enough to take off the jacket, and one may be tempted to lie down on the warm sand and take a nap.

At thirty feet, the foredune is less than imposing in height, but it is nevertheless an effective barrier to onshore winds that would otherwise sweep the central dunes as they sweep the beach. The absence of wind permits heated air near the ground to remain in place during the day, but it also permits pockets of cold air to develop at night. The result is significantly warmer days and colder nights behind the foredune than on the beach. Table 4.1 summarizes weekly maximum-

Period	Maximum temperature		Minimum temperature		Rainfall
	Strand	Behind foredune	Strand	Behind foredune	
Nov 20–Nov 28	61	74	39	28	1
Nov 28–Dec 4	64	66	34	30	0
Dec 4–Dec 11		no data			37
Dec 11–Dec 18		no data			53
Dec 18–Dec 27	61	64	48	54	125
Dec 27–Jan 5	61	68	32	23	0
Jan 5–Jan 10	56	61	50	32	99
Jan 10–Jan 17	61	59	52	48	105
Jan 17–Jan 24	61	61	54	48	155
Jan 24–Jan 31		no data			18
Jan 31–Feb 7	68	74	50	39	2
Feb 7–Feb 14	68	74	46	39	36
Feb 15–Feb 21	61	72	39	26	20
Feb 21–Mar 4	70	79	41	30	42
averages	63	68	45	37	

Table 4.1 Weekly maximum and minimum air temperatures (°F) 1 m above the ground on the strand and just behind the foredune.

minimum air temperatures, 1 m off the ground, during winter months on the strand and just behind the foredune. For the fifteen-week period shown, maximum temperatures averaged 5 °F warmer behind the foredune, but frosts occurred during five of the weeks behind the foredune, never on the strand.

Differences in air temperature, however, probably do not affect the success of seed germination, which takes place deep in the sand. We found, for the time period covered in Table 4.1, that temperature 5–10 cm below the soil surface did not differ significantly between strand and dune. This is the time and depth at which germination of sea rocket occurs. Sea rocket (Fig. 4.3) is common along the strand, but is never seen growing behind the foredune. Close examination of the soil surface between January and March will show many small sea rocket seedlings on the strand, but not behind the fore-dune. If these are dug up, the old dry fruit, split by the ger-minated seedling, can still be seen attached to the juncture of root and stem about 5–10 cm below the surface; the position of the fruit indicates that germination occurred at that depth. Obviously, temperature gradients are not the reason sea rocket is restricted to the beach: as far as the seed is concerned, gradi-ents don't exist, for temperature averaged 60°F in the soil on both the strand and in the dunes.

Incidentally, optimum germination temperature for sea rocket is not 60°F, even though all germination in nature oc-curs at that temperature. Using the temperature gradient bar (see Chapter 2 and Appendix B), we determined the effect of

Fig. 4.3   Sea rocket

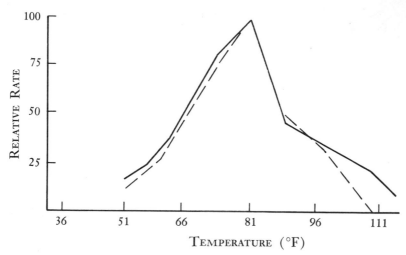

Fig. 4.4 Effect of temperature on germination (solid line) and seedling root growth (dashed line) of sea rocket.

temperature on sea rocket germination (Fig. 4.4). After four days' incubation in the dark, optimum temperature turned out to be about 82 °F. If the number of germinations at that optimum is given the relative value of 100, then Fig. 4.3 shows only one-fourth as many (25) will germinate at 60°F. (With time, germination at 60°F increases slightly: after fourteen days' incubation, germination at 60°F is up to 40% that at 82 °F.)

Sand at that depth does reach a temperature of around 80 °F at other times of the year. On a clear day in early October, when the air temperature was only 67 °F, temperature on the surface of a south-facing sand hummock was 105 °F, and temperature 10 cm below the surface was 85 °F. Even on the shaded, north-facing slope it was close to 70°F at that depth (Fig. 4.5). This sort of profile is probably typical of summer conditions on the Bodega Bay strand. Why doesn't sea rocket germinate at that time? Probably because summer is a dry time of year. Sea rocket responds just like the grassland lip

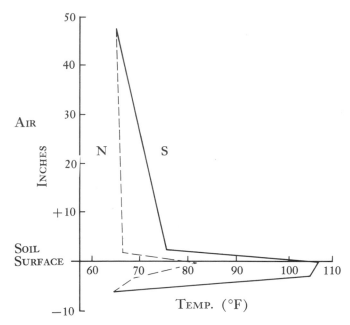

Fig. 4.5 Temperature at different distances above and below the soil surface on the north-facing (N) and south-facing (S) sides of a small dune hummock. Distance is given in inches. Clear October day, 2 P.M.

plant sea pink. Optimum soil temperature for germination of sea pink occurs in January (Figs. 2.10 and 2.12), but germination in the field occurs some time earlier, triggered by the onset of winter rains. Sea rocket germination is also tied to rainfall, but apparently it requires a greater amount of accumulated rainfall, for it germinated two to three months later than sea pink did.

Table 4.2 shows the timing and distribution of seedlings of sea rocket along the strand in 1970. The previous September, we had marked the centers of eight parent clumps of sea rocket by pounding tall poles into the ground. During subsequent winter storms, most of the parent clumps disappeared,

Clump	November 19	December 27	January 24	February 2	February 17	March 7	Average
1	0	0	1m = 71 2m = 4	1m = 45 2m = 0	1m = 80 2m = 2	1m = 97 2m = 2	1m = 73 2m = 2
2	0	0	1m = 4 2m = 2	1m = 6 2m = 1	0	1m = 4 2m = 0	1m = 4 2m = 1
3	0	0	0	0	1m = 1 2m = 0	1m = 0 2m = 2	1m = 0 2m = 1
4	0	0	0	0	1m = 1 2m = 0	1m = 1 2m = 0	1m = 1 2m = 0
5	0	0	1m = 8 2m = 0	1m = 11 2m = 0	1m = 29 2m = 0	1m = 49 2m = 7	1m = 24 2m = 2
6	0	0	1m — 35 2m = 3	1m = 9 2m = 0	1m = 52 2m = 0	1m = 38 2m = 11	1m = 34 2m = 4
7	0	0	0	0	0	1m = 4 2m = 0	1m = 1 2m = 0
8	0	0	1m = 1 2m = 0	1m = 5 2m = 0	1m = 141 2m = 0	1m = 200 2m = 41	1m = 87 2m = 10
Average	0	0	1m = 15 2m = 1	1m = 9 2m = 0	1m = 38 2m = 0	1m = 50 2m = 8	1m = 28 2m = 2

Table 4.2 Germination and establishment of sea rocket seedlings on the strand during the winter months of 1969–70. Seedlings were counted within circles of increasing radii from the center of parent sea rocket clumps: seedlings within 1 m radius, additional seedlings within 2 m and all others beyond 2 m (none seen).

but the poles continued to mark the places they had occupied. The poles were visited every week and the ground searched for seedlings. The first seedlings were seen on January 17, but the first full count was made one week later.

Aside from the striking phenomenon of most seedlings appearing within a radius of 1 m of the clump centers, the fact that stands out from the data in Table 4.2 is that seedling establishment fluctuated with rainfall. By referring back to Table 4.1, one can see that prior to November 20, rains were light and scattered. Three weeks of rain, averaging 1.3 inches per week, produced no seedlings by the end of December. Following a week-long drought, two more weeks of rain averaging 4 inches a week produced the first seedlings on January 17. After this first flush of germination, two subsequent dry weeks averaging less than 0.4 inches per week resulted in a decline in the mean number of seedlings about each marker from sixteen to nine. But after January 24, rainfall averaged nearly 1 inch a week, and the mean number of seedlings about each marker rose to fifty-eight.

Over 90% of all the seedlings arose within a radius of 1 m of the center of parent clumps. No seedling was seen farther than 2 m from the marked clumps, nor was any seen behind the foredune. Does this mean that conditions for germination are better near the parent clumps, or does it simply mean that the seeds are not distributed very far from the parents? We think the latter choice is the correct one, because we have artificially sown seeds of sea rocket elsewhere on the beach and behind the foredune, and many germinated. The dry fruit, with its enclosed seed, is relatively light, and one might expect it to be blown inland. But, judging by the absence of seedlings, this does not happen. Seeds must soon be buried by shifting sand, swept out to sea by storm waves, or trapped in the dense beach grass on the foredune.

Temperature gradients may not explain the distribution

of plants within the dunes, but they may partly explain the biological differences between dunes and grassland. In summer, the bare sand allows more heat energy to penetrate the ground than does the shaded grassland surface. Thus, within the dunes the temperature is higher in the root zone (5–10 cm below the surface, Fig. 4.6). During the two summer months shown in Fig. 4.6, dune topsoil averaged 11°F warmer than grassland topsoil. This much of a temperature difference may be enough to affect soil animals and microbes as well as plant roots.

SALINITY AND WIND

The decrease in wind speed behind the foredune affects salinity of air and soil, as well as temperature. Fig. 4.7 shows salinity of dune topsoil in October, at the end of the dry season. Salinity of the strand was more than four times as high as that just behind the foredune. Further inland, salinity remained constantly low. Salinity seems very unevenly distrib-

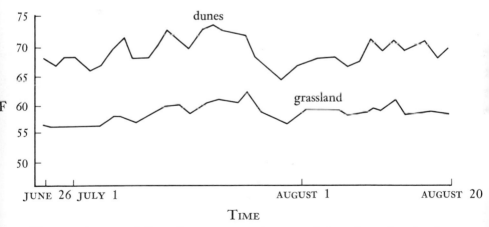

Fig. 4.6 Average daily soil temperature (5–10 cm below the soil surface) in the dunes and grassland during summer months of 1969. The dunes position was 1000 m inland from the strand, the grassland position was 310 m inland from the bluff lip.

uted over the dunes surface, however, and Fig. 4.7 should be
taken as indicative of a general pattern rather than of absolute salinities. On the strand, salinity in the root zone can
range from 450 to 1,800 ppm; at the time sea rocket seeds
germinate, it averages 1000 ppm. Behind the foredune, the
salinity ranges from 50 to 500 ppm, so in general the strand
is two to three times as saline.

The strand is also more saline than the grassland. Compare the January strand salinity of 1,000 ppm with the grassland lip soil salinity of 150 ppm at the same time (Fig. 2.15).
Germination is taking place at that time in both areas, but the
seeds germinating are completely different. Is the difference
between 1,000 and 150 ppm responsible for the species differences? Probably not. Optimum germination of sea pink oc-

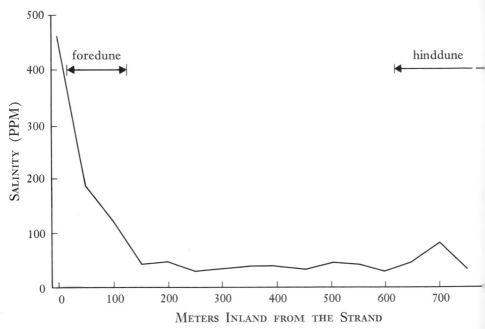

Fig. 4.7   Soil salinity (top 10 cm) at different distances inland from the strand.
Position 1 corresponds with the zone dominated by sea rocket. Samples collected
12 October 1968 at the end of the dry season.

curs at 0 ppm, but salt concentration of 1,000 ppm reduced germination only by 30% (see Fig. 2.17); therefore, in theory sea pink could germinate on the strand in January. But it doesn't. Sea rocket does germinate there, but Fig. 4.8 shows that considerable germination can occur at 100 ppm (sodium chloride). So sea rocket too could, but doesn't, germinate at the grassland lip in January. Other factors must be responsible.

Incidentally, the seeds of sea rocket which failed to germinate at high salinities (above 1% salt, Fig. 4.8) were not killed by the exposure; they were simply inhibited (temporarily) from germinating. When they were removed from the saline medium, washed with distilled water, and placed on a salt-free medium, a large number of them germinated. This behavior is shared by many beach plants, and it seems to be an

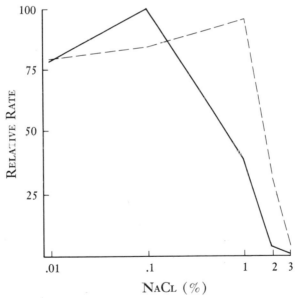

Fig. 4.8 Effect of sodium chloride concentration on germination (solid line) and seedling root growth (dashed line) of sea rocket.

adaptation to seed dispersal by ocean currents. Seeds of sea rocket, enclosed in the fruit, can float in sea water as long as eleven days. If they were swept to sea by storm waves, they could be carried many miles and deposited on another beach in that amount of time. Once on shore, winter rains leach the fruit of salt and permit germination. This seems to be how sea rocket has spread along the Pacific coast.

Sea rocket was first seen on the West Coast in 1935. Several plants were growing on a beach just north of San Francisco. They had probably grown from seeds accidentally carried into San Francisco Bay by ships from the East Coast of the United States, where sea rocket is found. Since that time, sea rocket has migrated north and south along the coast at an average rate of 33 miles per year. Today its range extends from the Queen Charlotte Islands off British Columbia to Cedros Island off Baja California, a distance of nearly 2000 miles. In contrast, the grassland weed bull thistle, whose seeds are dispersed on land, has travelled from its point of introduction at an annual rate of only 17 miles—and that is considered fast for a non-strand plant, most other introductions travelling less than 10 miles per year.

The salinity of water carried by the air may be more critical to the distribution of plants in the dunes than the salinity of the soil. To sample the gradient of salt spray, we set out petri dishes containing circles of filter paper at regular intervals back from the foredune. At the end of forty-eight hours, the papers were collected and the amount of salt deposited was determined (see Appendix B). During a two-day period in September, when wind speed averaged 11 mph, salt spray at the top of the foredune averaged 0.03 mg/cm^2/day, just behind the foredune it jumped to 0.08, but from there it continued to drop to a low of 0.01 at the start of the hinddune. Unfortunately, the dish left on the strand had been disturbed; based on measurements taken at other times, it probably was close to 1 mg/cm²/day on the strand. Salt spray did decline with increasing distance from shore except for a sharp rise just behind the foredune. The rise may have been caused by sudden decrease in wind speed resulting in a major settling out of spray particles. Is it coincidental that most lupine shrubs do not appear in the dunes any closer to shore than the base of the hinddune—and that salt spray deposition drops to 0.01 mg/cm²/day by that point? Large lupines were also absent near grassland ocean-facing lips, and did not occur until salt spray levels were below 0.01 mg/cm²/day (see Table 2.4).

Had the wind during this two-day period been blowing faster, more salt spray undoubtedly would also have been deposited. Working in dune vegetation along the East Coast some years ago, Stephen Boyce (1954) showed that salt spray level increased with wind speed. The increase was proportional except for a large jump in deposition when the wind speed was greater than 13 mph. Searching for an explanation, Boyce found that 13 mph is the low limit of force 4 winds on the Beaufort scale (force 4 = 13 to 19 mph). Force 4 winds also correspond with wind speed necessary to produce whitecaps, whose bursting foam bubbles catapult salt particles into the air (see Chapter 1).

Zonation of Organisms

PLANTS

Plant cover was measured along the entire 1,500-meter-long dune transect. The importance of each species to the community was estimated by calculating the percentage of the transect line it covered. Plant cover in each area is summarized in Table 4.3, where the transect is divided into three segments: strand and foredune, central dunes, and hinddune with protected area behind it.

The strand is nearly bare of plants, and the only two species which occur regularly on it do not appear in Table 4.3; they are introduced plants, sea rocket, *Cakile maritima*, and sea fig, *Mesembryanthemum chilense* (Fig. 4.9). Both

	Foredune	Central Dunes	Hinddune Area
Bare sand	22.0	80.4	35.4
Beach grass (*Ammophila arenaria*)	78.0	18.0	44.8
Lupine (*Lupinus arboreus*)		0.3	4.6
Coyote bush (*Baccharis pilularis*)			2.4
Ice plant or sea fig (*Mesembryanthemum chilense*)		0.5	
Evening primrose (*Camissonia cheiranthifolia*)		0.5	0.6
Lotus (*Lotus heermanii*)		0.3	
Silver beach weed (*Ambrosia chamissonis*)			0.2 0.2
Haplopappus ericoides			1.5
Thistle (*Cirsium occidentale*)			0.2
Sand verbena *Abronia latifolia*			0.8
Rush (*Juncus leseurii*)			8.2
Yarrow (*Achillea borealis*)			0.1
Cinquefoil (*Potentilla egedii*)			1.1
Moss species			0.1
total	100.0	100.0	100.0

Table 4.3   Plant cover (%) along three segments of the long dune transect.

have succulent leaves and stems with smooth, hairless surfaces, and low-lying growth forms—all of which possibly
adapt them to salinity in air and soil, and blowing sand. The
sea fig is a perennial and may occupy the same spot on the
beach for several years, but sea rocket is an annual and must
start from seed each year. (We have just seen, however, that
the position of the parent plant determines the position of
next year's seedlings.) Although many seedlings of these two
species are seen on the beach in winter, survival is low. By
the time sea rocket plants reach flowering and begin to set
seed in late summer, there may be only one clump of several
plants every 30 m, and sea fig is even less abundant than that.
(Actual counts of marked sea rocket seedlings showed 90%
mortality during a sixty-four-day period in the winter of
1970–71.)

The foredune is densely covered with a third introduced
species: beach grass, *Ammophila arenaria* (Fig. 4.10), originally from the shores of the Mediterranean. Seedlings have
been planted on the dunes area several times since the early
twentieth century to prevent sand erosion. It is an excellent
sand stabilizing plant because its extensive, fibrous root system
binds a great deal of sand, and because it spreads rapidly
through vegetation into open sand. The plants produce underground stems (rhizomes) which grow out in all directions,
sending up shoots at regular intervals. Beach grass is very

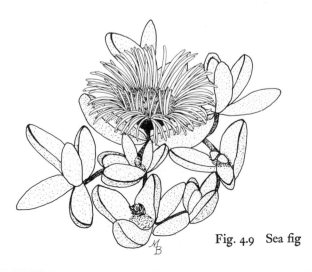

Fig. 4.9   Sea fig

abundant on the foredune, its sharp-pointed leaves making passage difficult; but in the central dunes it covers 30% of the area. (By chance, the long dunes transect passed through an area which may not have been planted to beach grass, and this is why Table 4.3 shows only 18% cover by beach grass.)

Several prostrate, perennial herbs are associated with beach grass in the central dunes: evening primrose, *Camissonia cheiranthifolia*, lotus, *Lotus heermanii*, and two species of sea fig. Commonly, entire patches of sea fig are black, dried, and dead; we do not know what killed them. Short leaves and flowering heads of sand blue grass, *Poa douglasii*, are occasional. Lupine shrubs are rare, but increase in frequency back toward the hinddune.

The hinddune and the protected area behind it show a great increase in plant diversity and in plant cover. Three shrubs are common: coyote bush or *Baccharis pilularis*, *Haplopappus ericoides*, and lupine. Several perennial herbs are scattered among the shrubs: silver beachweed, *Ambrosia chamissonis* (Fig. 4.11), beach dandelion, *Agoseris apargioides*, sand verbena, *Abronia latifolia* (Fig. 4.12), beach strawberry, *Fragaria chiloensis*, and the beautiful native thistle with

Fig. 4.10 Beach grass showing vegetative growth.

a cobwebby appearance, *Cirsium occidentale*. In contrast to the grassland, annuals are rare.

In depressions, where the water table may be at or near the surface, several plants characteristic of fresh-water marshes (Chapter 6) are present: a number of rush species (most commonly *Juncus leseurii*), mosses, and cinquefoil, *Potentilla egedii* (Fig. 4.13). Several grassland species manage to germinate in great numbers here, but few are able to reach maturity and flower by the time the sands dry out in early summer. Sow thistle and cudweed are two which do on occasion reach flowering in dune depressions.

As plant cover increases, the sandy soil becomes more stabilized. Organic matter falls more heavily to the surface and is decomposed; the decomposition products not only enrich the soil, but they improve its water retention properties and possibly affect pH, so that more demanding plants can live on it. Given time and freedom from disturbance, dunes may support shrubland or forest, as they do along many parts of the Pacific coast. The plant ecologist M. L. Kumler (1969) has recently documented the path of community succession, from sand dunes to forest, on the Oregon coast.

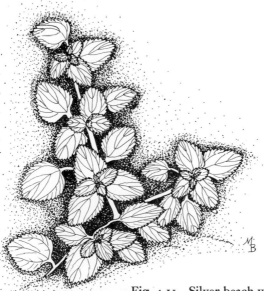

Fig. 4.11   Silver beach weed

Kumler visualized a pioneer community on exposed coast (beach) as in time changing the habitat, and giving way to a mixed grass-herb community (yarrow, pearly everlasting, velvet grass, sheep sorrel, strawberry, beach grass, and others); then this community in time giving way to one dominated by shrubs (bearberry, blueberry, salal, rhododendron, wax myrtle, and bracken fern), and finally this community ultimately giving way to a coniferous forest of lodgepole pine, sitka spruce, and mountain hemlock. Although the forest species are not present now at Bodega, many of the herbs and shrubs mentioned above are. There is, then, potential for the protected dunes at Bodega to become much shrubbier, given protection from disturbance.

Disturbance removes plant cover and sets the sequence of community succession in reverse. A small open area, created by a horse trail or an excavation for fill or a dune-buggy track, can be widened by wind erosion, exposing the roots of plants along the border, or by these plants being buried in moving sand. The once-small open area turns into a large, growing one: a blowout. The amount of sand removed from a blowout can best be seen from the size of isolated hillocks within it. These miniature mountains of sand have been held in place by roots of occasional plants that somehow managed to remain growing as the blowout engulfed them. The top of the hillock represents the previous dune surface. Wayne Williams (1972, 3) of California State Polytechnic College re-

Fig. 4.12   Sand verbena

cently documented the havoc that dune buggies created in Morro Bay State Park, California. Once-stable dunes, covered with a shrubby community of buckwheat, *Eriogonum stachadifolium*, live forever, *Dudleya farinosa*, *Haplopappus* and *Corethrogyne*, reverted to moving dunes with scanty pioneer plants (sea rocket, sea fig, sand verbena, evening primrose, others). Exposed sand has accelerated the deposition of sand in the harbor and created dredging problems. Williams estimated that the cost of disturbance could be equated with the cost to the town of Morro Bay of more frequent dredging.

What restricts pioneer plants, like sea rocket, to the severe beach habitat? Why don't they occur in more stable communities, such as the protected hinddune area or the grassland? We think there is more than one factor responsible, and we would like to use sea rocket as an example.

In December of 1969 we planted seeds of sea rocket on the strand and just behind the foredune. They germinated the following January, when normal seedlings did (Table 4.2); germination was higher, and seedling growth faster, behind the foredune. But by April, all the seedlings behind the foredune had disappeared, while a few still remained on the strand. Some of the seedlings which disappeared had clearly been grazed: the shoot material was gone, but the stump end of the stem could still be seen at the sand surface. Differential

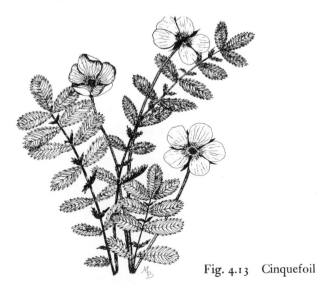

Fig. 4.13 Cinquefoil

grazing pressure, between the strand and dunes, may be one of the factors restricting sea rocket to the strand. (On a previous page, we have already seen that another factor may be poor seed dispersal inland.)

Competition for light, rather than grazing, may eliminate sea rocket from the grassland. We sowed several hundred seeds of sea rocket in March in four different seedbeds in the field. One seedbed, which served as the control, was the natural strand. Another was undisturbed grassland soil, with all the other plants left in place. A third was disturbed grassland soil, with all grassland plants removed. A fourth was also weeded of grassland plants, but in addition the top 5 cm of soil was replaced with sand from the beach. After sowing the seeds, the plots were revisited every week to check for germination and plant development.

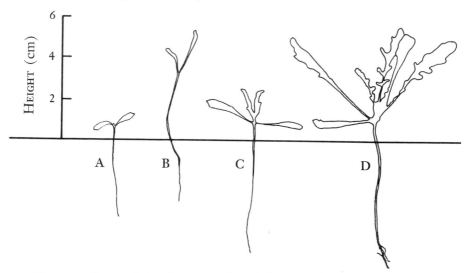

Fig. 4.14 Stage of development of typical sea rocket seedlings six weeks after being sown in four different habitats: A, normal strand; B, normal unweeded grasslands; C, weeded grassland with strand sand instead of grassland soil; and D, weeded grassland with grassland soil. The horizontal line represents ground level; most of the root material, however, was not collected.

Germination rate, after four weeks' time, was roughly equal for all four treatments. But by the sixth week, seedling survival and plant growth were markedly different. Fig. 4.14 shows a representative seedling for each plot at that time. Seedlings from the strand were quite small, with only succulent, red-tinged seed leaves projecting at the surface. By far the best growth and survival rates were exhibited by seedlings in the grassland soil weeded of all competing plants. Growth in the undisturbed grassland plot was poor; seedlings were spindly, weak, pale, and leaves were not well developed —all symptoms of low light intensity. Two months later, none of these etiolated plants remained. In contrast, three of the plants in the weeded grassland plot flowered and produced viable seeds by late summer.

Laboratory tests have confirmed the hypothesis that sea rocket seedlings need a high light intensity. Seedlings of similar age and stage of development were transplanted to flats of sand and grown under different light intensities for four weeks. Light intensity was measured in units called footcandles (see Appendix B): on a clear day, light intensity at the ground might be 7,000 ft-c; in the shade of a tree it might

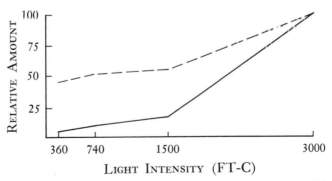

Fig. 4.15 Effect of light intensity on plant weight (solid line) and leaf development (dashed line) of seedlings of sea rocket.

be less than 500 ft-c; in among the crowded plants of a grass-land it might be less than 100 ft-c. The levels we chose to experiment with were 3,000, 1,500, 740, and 360 ft-c. Fig. 4.15 shows that growth of sea rocket seedlings was significantly reduced when light intensity fell below 3,000 ft-c. Compared to plants grown in 3,000 ft-c, plants grown in 360 ft-c had a plant dry weight reduced by 90% and leaf development reduced by 45%.

ANIMALS

The strand, or sandy intertidal zone, appears desolate by comparison with the nearby rocks, and not many kinds of animals are found there. However, a number of species build up large populations—as often happens in areas of low species diversity. And, if one knows where to look, one can find a larger number of animals than immediately strike the eye. Part of the problem facing a collector here is that many of the animals burrow, or they hide under wood and seaweed cast up on the shore. That the strand is not so underpopulated as one might think is readily apparent from the fact that many birds seek food here. These birds, mainly members of the sandpiper family, will be treated in the next chapter.

Sand provides a moist but unstable substrate, and one that heats up at the surface when the tide is out. It tends to abrade the animals, especially when moved by wind and waves. The instability becomes particularly great on an area like Salmon Creek Beach, where the wind and waves come in at full force. On the spit near the breakwater, the forces are diminished somewhat, and a number of more delicate animals are found, but we shall restrict this discussion to the more extreme habitat.

Among the most abundant animals of the intertidal zone is the common "sand crab" or "mole crab," *Emerita analoga.* Very streamlined, it burrows rapidly, backing into the sand

and sticking its head into the water. A filter feeder, it has modified appendages which serve to extract suspended food from the water. They move up and down with the tides, and the smaller individuals are thought to occupy a somewhat higher position. This species has lately been discovered to be a "protandrous hermaphrodite"—the same individual first passes through a male stage, then becomes a female. The smaller males may be seen pursuing the females over the surface of the sand at night. If one lifts up the tail of a large individual, he can often see a cluster of yellow-orange eggs (see Cubit 1969).

Above the waves are a number of small animals, including some which migrate when the tide is in. The place to look is under boards and among and around heaps of rotting seaweed. The latter provide moisture and protection, and food as well. The animals here are all arthropods, whose exoskeleton helps provide some barrier to desiccation. Insects, which are invaders from the land, flourish about the heaps of seaweed, as can be seen from the swarm of flies which hover over them, and from the several species of beetles among and beneath. They are joined by some crustaceans, which have moved in from the sea: amphipods or beach-hoppers, such as *Orchestoidea californiana*, and some isopods or "pill bugs," such as *Alloniscus perconvexus*.

Likewise in the dunes one finds animals low in both diversity and density; and many are restricted to the protected area behind the hinddune or to depressions which are seasonally wet. Deer mice, for instance, are restricted in burrowing activity to the stabilized sand beneath dense clumps of beach grass or lupine shrubs. Voles and shrews are found only in areas heavily covered with beach grass.

Jack rabbits and deer roam the entire dunes area. The small, round jack rabbit scats are very commonly seen, and patches of matted beach grass indicate resting places for deer.

A complex insect community, which we have not examined in detail, exists at and below the sand surface. It includes herbivores (aphids, marsh moth caterpillar, some ants), carnivores (aphid wasps, ant lions, cricket wasps), and some that change their food habits with age, like the hover fly which preys on aphids as a larva but lives on flower nectar as an adult. Tracks of weasel and raccoon are common throughout the dunes.

The dunes area does not support a large bird fauna, except around the seasonally wet depressions, and these species will be discussed in Chapter 6. The temporary ponds also attract a number of animals who reside in the grassland: badger, skunk, fox, raccoon, and weasel. The marsh hawk nests in the dunes and does pass over it, hunting prey, but spends more time over the fresh-water marsh and grassland. Foxes, raccoons, and weasels are common throughout the dunes.

5

The Mudflat Community and its Suburbs

Background

In attempting to discuss the ecology of the mudflat community and Bodega Harbor, we find ourselves overwhelmed with
data and new discoveries. Comparable habitats on the West
Coast of North America have already been extensively investigated. For example, Warme (1971) worked on the relatively
undisturbed Mugu Lagoon in southern California, MacGinitie
(1935) on Elkhorn Slough north of Monterey, Macdonald
(1969) on salt marsh biotas all along the coast, and various
workers, from the Pacific Marine Station, on Tomales Bay
just a few miles south of Bodega Head.

Fig. 5.1a   Mudflats at high tide.

Fig. 5.1b Mudflats at low tide.

To what has already been published, we here add some of our own studies on the Bodega Harbor salt marsh. In addition, we can draw upon a recent survey in which one of us (M. T. Ghiselin) has participated. During the spring quarters of 1970 and 1971, faculty and students from both the Berkeley and the Davis campuses of the University of California, representing a broad variety of disciplines, studied the ecology of the harbor ecosystem. About one hundred persons were involved. A large study area next to the Marine Laboratory property has been mapped, and sampled both quantitatively and qualitatively. The physical makeup of the habitat, the community structure, and the biology of the dominant organisms have all been studied in great detail, and the results have been placed in an evolutionary perspective by comparisons with fossil assemblages. We are dealing, therefore, with one of the best understood ecosystems of its kind in the world. Professor Hamner of Davis is now supervising the preparation of a monograph based on this research (Hamner MS). We shall here present a summary and preliminary report, leaving the detailed documentation for the more technical publication.

Description of the Habitat

The marine habitat located within Bodega Harbor may be spoken of as a "mudflat" fringed, in many places, with salt marsh (Fig. 5.1a and b). One could treat the marsh either as a modified portion of the mudflat itself, or else as an ecotone, joining the terrestrial habitat with the marine one. The situation here is somewhat artificial, since the harbor is kept open at the breakwater, and a channel has been dredged for boats. This keeps the salinity and other properties of the water more like those on the outside, and provides a place where forms from somewhat deeper water can live.

To call the major habitat a "mudflat" is a bit misleading.

For some places "sandflat" would be a better term. Sand continually blows into the harbor from the dunes area, and the harbor is filling up in spite of repeated dredging. Toward the channel—away from the source of the sand, that is—grain size becomes a bit smaller, and there is more clay, silt and particulate organic matter. The sediment, however, is fairly uniform, especially at any one place where burrowing animals mix it up a great deal.

The harbor itself may instructively be viewed as a shallow, partly-isolated portion of the marine environment. Only a modest amount of fresh water comes into it, from a few small streams, runoff from rains and ground water. Furthermore, the harbor gets flushed out and refilled frequently. On a good low tide, just about everything drains all the way to the channel, and even at neap tides much of the water gets exchanged every day. Fairly hardy intertidal and subtidal forms occurring on the outer coast live on the scattered rocks and pilings around the harbor. Since rain water may flow over the surface of the mud, the animals living below are somewhat protected from the lessened salinity above. The large surface area does result in warmer water in the harbor, and the effect is accentuated during neap tides, when flushing is only partial.

Some further effects of the flushing regime should be mentioned. First, the large area covered by the shallow layer of water brought in by the tide means that a fairly small volume of water from the open sea is available per mudflat inhabitant. And when this water warms up, it is too warm for the open-ocean phytoplankton; consequently, the amount of energy coming in from the water itself should be less than it is on the open coast. Mudflats, however, are thought to be extremely productive. The surface and subsurface mud contains an abundant population of small plants, mainly diatoms. At lower levels there develop extensive beds of the "eelgrass"

Zostera marina, which, like surf grass, is a flowering plant. Somewhat higher, especially among pools, such algae as *Ulva* and *Enteromorpha* are prominent in spring and summer. The marsh as well contributes much food. A good educated guess is that the mudflat gives more enegry to the rest of the world ecosystem than it receives. Another important point has to do with the effect of the narrow opening at the breakwater. It dampens the movement of the tides, and virtually eliminates waves from outside. Tides within the harbor occur some minutes later than they do on the open coast (high tides about sixty minutes, low tides about seventy minutes later), and their amplitude is diminished. By the time the tide turns, the level within has not reached the level without. Consequently, the intertidal zones become somewhat compressed. Their boundaries, already rendered difficult to see because of the very gradual slope, become even more obscure.

When the tide comes in, it brings food to the mudflat animals, and also brings in animals that feed on both animals and plants. When it goes out, the inhabitants are exposed to another group of predators and herbivores, and to the drying action of the air. Both of these influences can be somewhat ameliorated by burrowing. Even diatoms have been shown to migrate up and down in the mud. The water also brings another necessity of life, oxygen, and when no water is available the burrowing animals must be able to get along under conditions of reduced oxygen availability, often for protracted periods. Burrows, to be sure, may bring in oxygen from the air, and some mudflat animals can breathe a certain amount of air. But when the tide is out, many of the animals have to do without an extrinsic oxygen supply, and this necessitates one or another adaptation or a combination of these: dormancy, relying on stored oxygen, or going into oxygen debt. Hollibaugh (Hamner MS), using the standard Winkler technique, found that below 2 cm depth in the mudflat the oxygen level never was more than 0.05 ml of oxygen per liter of water—the limit of resolution for this method. The mud is dark in many places, indicating anaerobic conditions.

Zonation of Organisms and of Communities

Zonation being less obvious on mudflats than on rocky shores, we have far fewer, and far less satisfactory, classification schemes for this habitat. One of the more interesting results of the surveys carried out in 1970 and 1971 has been that a distinct pattern does in fact emerge, and that it parallels, in a way, the relationships already described for the rocky intertidal. As before, we find that the natural economy changes from zone to zone. In the present chapter we shall be able to explain what mechanism it is that determines this particular relationship. If one analyzes the biota at various levels, rather than along lines suggested by certain Russian ecologists (review in Walker 1972), one finds that the organisms in each zone have a characteristic manner of obtaining food. And this fact can, in turn, be related to the amount of time submerged.

Following Ricketts and Calvin, we shall here number the zones from 1 to 4, and draw the lines at about the same level as did they. The correspondence, however, is only approximate, especially since we find it so hard to demarcate the upper two zones. Zone 1 will cover the areas above mean high tide. Roughly it means the salt marsh, but this assemblage in places extends downward, and outside the marsh the zone 2 organisms live at the same level. It corresponds to zone 1 on the outer coast. Viewed in the context of our "trophic analysis," we would say that the zone is dominated by the terrestrial plants which have adapted to marine conditions, and by marine animals which receive some protection owing to their shelter and stabilizing influence on the substrate. Where marsh at this level does not occur, we find what look like zone 2 animals, but maybe these are better considered zone 1 forms which cannot compete in the marsh.

Zone 2 extends downward to about the mean higher low water, as in Ricketts and Calvin's scheme. The surface, except for a film of diatoms, looks very lifeless. The characteristic animals which live here are "deposit feeders." They eat materials off the surface, or extract it from the mud itself. Many actually ingest the sand, which is covered with a film of bacteria. Again, zone 3 extends to mean lower low water, as it does in other intertidal zonation systems. It is dominated by suspension feeders rather than by deposit feeders. That is to say, the animals here extract food suspended in the water, rather than deposited in or on the substrate. Zone 4 naturally encompasses the remaining intertidal. This zone is rather more difficult to characterize in terms of a trophic analysis. However, different groups of suspension feeders become dominant, eelgrass appears, and a variety of predators, especially those living above the surface of the mud, add to the diversity.

The explanation for this zonation pattern is basically the same as that proposed for rocky intertidal habitats in Chapter 3. On an average day, distinct, rapid changes occur in the time of submergence at the levels of mean lower water, mean higher low water and mean lower high water. From mean lower high water upward, we have periods of occasional flooding—the ground is kept fairly moist, but subsurface feeders are at a disadvantage, one which rapidly gets accentuated toward higher elevations. Below (zone 2) the tide is usually out, but the area is covered at regular intervals. Hence, the suspension feeders would have little time for feeding. In zone 3 the tide is in most of the time, and it becomes profitable to feed upon suspended matter. At even lower levels, the average conditions become virtually like those in the subtidal, and hardy forms can endure occasional exposure while others can migrate. Feeding relationships, and not physiological stress, thus largely determine the zonational pattern. But only when we focus our attention upon the burrowing forms, which can avoid the physiological stresses, does it become obvious that this is so.

ZONE I: PLANTS

Salt marsh fringes much of the harbor, but its actual area is not very extensive, partly because man has altered the conditions. On the west side of the harbor on the marine preserve property, it is less than 60 m wide. This land corresponds closely with the area exposed at mean high tide, but covered by the highest high tide. The marsh vegetation begins abruptly along the shore at the mean high tide level, extending inland on gradually rising land. When the soil surface is wet, blue-green algae grow there abundantly; they dry to a thin crust during periods of lengthy exposure in summer. It is dominated by low, perennial herbs that are dull brown during their winter dormancy, but green and vigorous in summer. Many of the species become tinged with red.

One might compare the salt marsh with the upper reaches of the rocky intertidal zone. The substrate differs, and the animals and plants there need not contend with waves or large amounts of salt spray. And at low tide, the salt-water marsh is a wet habitat, for the sand retains a great deal of moisture. The water, however, tends to vary somewhat in salt content, and one might want to compare the conditions here with those in high-level tide pools. If one digs down through the sand to a depth of eight to ten inches, the bottom of the hole begins to fill with water, indicating that the water table has been reached. At high tide, the water table rises even closer to the surface. But is this water sea water, or is it perhaps ground water seeping in from the nearby land? The answer depends upon the season, the time of day, and where in the marsh one happens to have dug.

There are two conflicting bodies of water opposing each other in the marsh: salt water from the harbor and fresh water from the land. The entire Head acts like a sponge during the rainy season, trapping water beneath the surface in a network of cracks and channels and small reservoirs, and permitting it to emerge later in seeps at the edge of the Head. A number of these seeps empty into the marsh on the west side of the harbor. Beneath the surface of the marsh, these two bodies of water intermingle, mix, or sit on top of each other. Depending on the season, one or the other has the dominant influence.

We measured the ground water salinity below the center of the marsh periodically through the 1969–70 period shown in Fig. 5.2. Salinity fluctuated from 1.1% in late summer to 0.1% in spring. Ocean water has a salinity of about 3.5%, while water of 0.1% doesn't even taste brackish; it is only

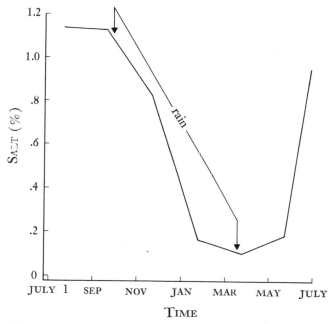

Fig. 5.2 Salinity of the ground water in the center of the salt-water marsh from 1 July 1969 to 1 July 1970. The first and last rains of that period are marked with arrows.

three times as salty as tap water. The ten-fold drop in salinity occurred between October and April, which was the rainy season. When the rains ended in late March, the amount of fresh water entering the marsh from seeps slacked off, and the influence of harbor water became more and more dominant. Water salinity increased to its previous high by the end of June. Average ground water salinity during the entire year was 0.5%.

The curve shown in Fig. 5.2 fails to show the many daily fluctuations. When the tide is in, the volume of sea water increases beneath the marsh, and the ground water becomes more saline. When the tide is out, the balance shifts toward fresh water, and the ground water becomes less saline. To document these changes, we established three replicate study transects, each 50 m long, beginning at the edge of the vegetation and extending inland nearly to the end of the marsh (see Fig. 1.2). Holes were dug down to the water table every ten m along each transect. During periods of at least twenty-four hours, during which the tide rose and fell several times, water samples were collected from each hole and analyzed for salt content with a salinity meter (see Appendix B).

Fig. 5.3 shows the results of one thirty-six-hour run. The results are somewhat confusing, but they do show that the salinity of the ground water can more than double in the course of a day. Changes in salinity were strangely out of phase with changes in tide level. The peaks and troughs in the ground water salinity correspond to those in tidal level, but a lag of about three hours is obvious. At the moment we aren't sure exactly what is going on here, but it should be obvious that water masses passing through sand can hardly be expected to move as rapidly as would a layer moving across the surface of the mudflat. The same figure also shows that salinity of the ground water decreased inland, away from the harbor. Salinity 5 m from the edge of the border on the har-

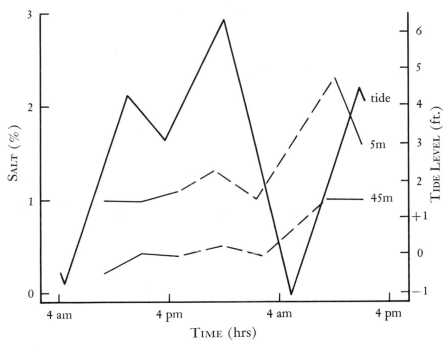

Fig. 5.3   Salinity of ground water at two positions in the salt-water marsh (5 and 45 m back from the mean high tide line) from 4 A.M. 30 June, to 4 P.M., July 1, 1970. The dashed portions of the lines are estimates, based on sampling done at other dates. Tidal fluctuations are also shown for comparative purposes.

bor side averaged about twice that at 45 m inland. This is to be expected, for the salt is coming in from the sea and the fresh water is seeping in from the land.

We studied the zonation of plants within the marsh. Even a beginner can easily familiarize himself with the species of plants in a salt marsh. He can almost qualify as an authority on a world-wide scale with modest effort. It is true that Herbert Mason's *A Flora of the Marshes of California* is almost 900 pages long, but not much of it is devoted to salt marsh flora: he included species found in twenty-six different kinds of marshes, ranging from open salt lakes, to seasonally wet fresh-water marshes, to hillside bogs, and even to strand. If one visits salt marshes around the world (not exactly a dream vacation), he will see, again and again, representatives of the same handful of genera. The Bodega Head salt marsh is no exception, but it does have some local genera of its own as well.

Of the fifteen vascular marsh plant species at Bodega, almost all are perennials which reproduce profusely by rhizomes. The pattern of seasonal activity for these plants is just

Species	\multicolumn{11}{c}{Distance (meters) from mean high tide}
	0	5	10	15	20	25	30	35	40	45	50
Salt grass, Distichlis spicata	63	26	16	7	28	32	22	27	16	31	27
Pickleweed, Salicornia virginica	21	70	80	83	65	47	36	4	2	12	20
Bulrush, Scirpus americanus			3	4	4	11	12	7	5	5	6
Arrow grass, Triglochin maritima						1		2	3		
Jaumia, Jaumia carnosa							22	60	74	45	31
Annual bulrush, Scirpus koilolepis										7	6
Bird's beak, Cordylanthus maritimus											7
Plantain, Plantago lanceolata											1
Bare ground	16	4	1	6	3	7	8				2

Table 5.1 Average ground cover (%) for all plants at different distances from mean high tide line in the salt marsh. Averages of three quadrats at each distance. Sampling was done in June 1970.

the reverse of that for grassland plants: they are dormant in winter and spring, and grow actively in late summer. The period of active growth corresponds with the period of high ground water salinity (Fig. 5.2), but, as will be shown later in this chapter, this is not a cause-and-effect relationship.

Distribution of these fifteen species is anything but homogeneous or random. Rather, groups of them reach peak abundance at different distances from the mean high tide level. To document this, we sampled plant cover along the three transects mentioned earlier. A wire frame encompassing an area of ¼ square meter was placed on the ground every 5 m. At each locality sampled, we estimated the amount of ground within the frame covered by each species. An average was computed for the three replicates at each location. The results are summarized in Table 5.1.

Close to the mean high tide level, only two species were found: salt grass, *Distichlis spicata* (Fig. 5.4), and pickleweed, *Salicornia virginica* (Fig. 5.4). The salt grass species is widely distributed in North and South America. The genus *Salicornia* occurs throughout the world. Both tolerate the most saline part of the marsh, but they have adapted to it in different ways: one secretes excess salt, the other retains it. Thus, small white dots can be seen on the leaves of the salt grass: the leaf tissue has exuded drops of liquid which dry out and leave a small deposit of salts. Pickleweed absorbs the salts and retains them. Its succulent tissue has some internal tolerance to high salt content. If you bite through the leafless, jointed stem, the tissue tastes noticeably salty. As shown in Table 5.1, salt grass is most abundant at the mean high tide level. It is the only species found uniformly throughout the entire marsh, and it contributed about 30% to ground cover along the entire transect. Pickleweed extended the length of the transect, but it declined in abundance inland, from a peak just back of mean high tide.

At the point where pickleweed begin to decline, other species appeared for the first time: a tall bulrush with a three-angled stem, *Scirpus americanus*, sea fig, *Mesembryanthemum chilense*, and arrow grass, *Triglochin maritima;* all of these are perennials. In the middle of the marsh the large bulrush in turn began to decline in abundance and other species appeared: *Jaumia carnosa* (Fig. 5.4); an annual bulrush that looks like fine grass, *Scirpus koilolepis;* and the only shrub in the marsh, the salt-secreting *Frankenia grandifolia*.

A number of species are restricted to the highest, most inland portions of the marsh which are rarely submerged by high tides. Some of these did not appear in Table 5.1, including two species of dodder, parasitic flowering plants. They lack chlorophyll and cannot make their own food. Instead, their weak, yellow or orange stems twine about leaves of *Jaumia* and pickleweed, and they send root-like organs into the host plants which extract nutrients from the tissue. Our

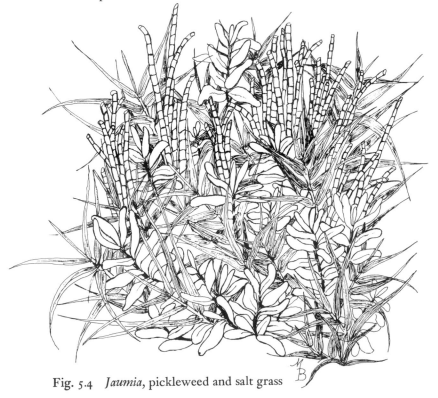

Fig. 5.4   *Jaumia*, pickleweed and salt grass

observations thus far indicate that one species of dodder parasitizes *Jaumia*, the other picklweed (see notes in checklist,
Appendix A).

Other inland plants include bird's beak, with purple-
tinged foliage, and *Atriplex patula hastata*, with scurfy-white,
deltoid-shaped leaves. Occasionally, plants more characteristic
of other areas also occur at the inland marsh fringe: a plantain
common on footpaths (see Chapter 7); sea rocket of the
beach (see Chapter 4) and cinquefoil and brass buttons of
fresh-water wet areas.

The Bodega Head salt marsh vegetation has a floral composition much like that of other marshes from Santa Barbara
north to Washington. Coastal salt marshes in southern California, however, have a greater diversity of species and exhibit more striking zonation. These marshes have definite
lower and upper regions. Macdonald (1969) found that the
lower marsh is dominated by tall cord grass (*Spartina foliosa*)
and a few other species not present at Bodega; it is submerged
for more than six hours at a time, at least once every fifteen
days. The upper marsh has a number of zones, one dominated
by pickleweed, one dominated by a complex community similar to the one at Bodega Head, and sometimes a grass-domi-
nated community further inland: it is submerged for less than
six hours at a time, no more often than once every several
weeks. The Bodega Harbor marsh evidently corresponds only
to the upper zone of southern California: an equivalent lower
marsh is for some reason absent. The reasons for such diversity patterns remain unclear. Yet it is intriguing to note that
the salt marsh diversity gradient, in which terrestrial plants
become increasingly depauperate, meets a comparable gradient in which fewer and fewer marine animals occur. It is clear
from Table 5.1 that plant diversity increased inland, just as
it did in the grassland and dunes. In the latter habitat we
enumerated many likely causes for this plant zonation. In the

salt marsh, only two are apparent: salinity of the ground water, and frequency of inundation. Perhaps the latter factor involves an indirect effect on the animals, but we shall defer this topic until later.

One hypothesis that might explain the correlation between salinity and plant distribution is that the plants which live closer to the shore grow better when the concentration of salt in the soil is high. To see if this is so, we grew salt marsh plants in the greenhouse in different dilutions of water, comparing the response of salt grass and pickleweed from lower elevations with that of *Jaumia*, *Frankenia* and sea fig from higher levels. Rhizomes of these plants were collected in February, while they were still dormant. The rhizomes for each species were apportioned equally into four lots, and each lot was placed in a 5 cm deep tray and covered with sterile sand. Holes were punched in the bottoms, and the trays of sand were irrigated by placing them in larger trays filled with tap water. The rhizomes were maintained this way for a month in a greenhouse, during which time the rhizomes broke dormancy. Incidentally, this shows that temperature is more important than salinity in breaking dormancy, since the greenhouse averaged about 20°F warmer than at Bodega, where salt marsh plants remained dormant. However, experiments with special controls would have to be done before we could rigorously infer that temperature differences really account for what we observed.

At the end of the month, the rhizomes of all the plants had broken dormancy, and several shoots had penetrated 1 to 10 cm above the sand in each tray. The length of each main and lateral shoot was measured, and the trays were subirrigated for four weeks with different dilutions of sea water containing some nutrient solution. The sea water was collected in Bodega Harbor, and was not sterilized or filtered. Full-strength, its salinity measured 2.8%, and the dilutions ranged from 2.2% to only tap water with nutrient solution. At the end of this experimental period, the length of the main and lateral shoots was again measured, and growth was calculated for each tray as percentage increase over the total length. As shown in Table 5.2, each species grew best at the lowest salinity (0.1%) and increasingly less as salinity increased. None of the species survived more than two weeks when sub-irrigated with water of 2.2% salts; but this isn't surprising considering that the highest salinity to which the plants are exposed under natural conditions is only 1.1%. (See Fig. 5.2.)

The relative tolerance of the five species to 1.1% salinity did not correlate with their positions on the marsh. *Frankenia*, which doesn't occur within 40 m of mean high tide line, grew much faster at that salinity than did salt grass, which grows right up to the water's edge. Also, *Frankenia* and *Jaumia* have the same marsh distribution pattern, but *Frankenia* proved much more tolerant of high salinity.

	Water salinity (%)			
Species	0.1	0.5	1.1	2.2
Salt grass, Distichlis spicata	2,744	345	90	0
Pickleweed, Salicornia virginica	2,021	1,075	791	0
Ice plant, Mesembryanthemum chilense	378	85	13	0
Frankenia, Frankenia grandifolia	3,367	644	318	0
Jaumia, Jaumia carnosa	285	231	0	0

Table 5.2   Growth of five salt marsh species during a four-week period when watered with different dilutions of sea water, in flats of sand in a greenhouse. Growth is expressed as percentage increase in total stem length during the experiment.

Surprisingly, all five species also grew better in soil drier than it usually is in their normal habitat. Duplicate trays of rhizomes and sand were not subirrigated, but rather watered from above once a day with tap water. As a result, the soil was not constantly saturated as it was in the first set of trays. Plants in these trays grew faster and flowered in greater profusion than those in saturated soil. This result indicates that marsh plants must contend with poor aeration as well as high salinity.

These experiments show that all five species could probably grow quite well in a fresh-water marsh or on relatively dry soil, such as in grassland. But under natural conditions, as at Bodega Head, they are in fact restricted to the salt marsh, and have never been found in any other habitat. Why? One possible answer is that the salt marsh plants are at some kind of competitive disadvantage outside of the habitat in which they ordinarily live. The mechanisms of such competitive interactions are but poorly understood, and it is not clear what resources may be in short supply. Perhaps the species which inhabit nonsaline soils have faster root and shoot growth rates than do the marsh plants, and reduce light penetration or soil moisture below acceptable limits. A similar phenomenon was seen to restrict sea rocket from the grassland. Conceivably the plants compete for space in a fashion like that known for barnacles. We simply don't know. But literature reviews by Barbour (1970) and Chapman (1964) make it clear that probably not a single marsh species requires high salinity to grow; most of them germinate only in the wet season, when salinity is low. At any rate, it appears that salt marsh species live in a physiologically sub-optimal habitat, and succeed because they alone can cope with the environmental difficulties.

ZONE I: ANIMALS

Animal life within the marsh easily escapes notice, for many of the inhabitants are small or otherwise inconspicuous.

Others should be treated as visitors or migrants from the land, namely some birds, mammals and insects. We shall defer our treatment of the terrestrial component of the marsh fauna, therefore, until later in the chapter, after we have discussed the resident plants and animals of each zone.

Among the roots of *Salicornia* and *Distichlis* one finds burrows of the common high-level shore-crabs. Outer coast forms are found here, both a good number of *Pachygrapsus crassipes* and an occasional *Hemigrapsus nudus:* the latter, however, is largely replaced by its relative *H. oregonensis*, a form which seems to flourish more in and around muddy situations in general. An amphipod crustacean, *Orchestoidea californica*, is common at this level but seems to prefer sand. Of the common small gastropods, one species, *Assiminea californica*, belongs to a gill-bearing group; another, *Phytia myosotis*, has a lung. In the *Salicornia* beds and nearby, the mud supports a luxuriant growth of a yellow-green alga, *Vaucheria*. A little green sea slug of the order Sacoglossa, *Alderia*, feeds upon its small filaments, and evidently eats nothing else.

At places where the marsh is not developed, animal life on the sandy to muddy substrate tends to be rather sparsely developed. However, forms perhaps better considered typical of zone 2 do live at this level. It seems reasonable to infer that some kind of competitive interaction may be going on, with the rooted plants tending to stabilize the substrate and exclude burrowers, and, conversely, burrowers creating unstable conditions making it hard for the plants to establish themselves. Rhoades and Young (1970) have adduced evidence in support of some ideas along these lines, applying them to subtidal communities. And it has been possible for us to document the same phenomenon for the "phoronid beds" discussed below (Hamner MS).

ZONE 2

This zone, like the third, has very little growing on its surface. Most of the action goes on down below. One may discern, as we have mentioned, a change in the trophic pattern between zones 2 and 3, but this is obscured to some extent by the fact that the two ways of making a living are practiced at both levels and elsewhere too, so that the difference is one of emphasis or degree. A comparable phenomenon was remarked upon in the chapter on rocky shores, where deposit feeders are virtually absent. Nonetheless, we saw that filter- (suspension) feeding groups increased as one went down, though some filtration was observed at all save the highest levels. The mudflat too supports a few wide-ranging suspension feeders that occur all across the mudflat. One of these is a tiny clam, *Transenella tantilla*. It lives on or just below the surface of the sand, and filters food from the water with its gills. Its ability to use water in the surface area when the tide is out might help to explain its success at higher levels, but this aspect of its biology has not been investigated. An hermaphrodite which bears its young alive, it builds up enormous populations which form a very important item in the diets of birds, fishes and other animals. Recently, two "morphs" of this "species" have been discovered which may prove to be different species. One of these is more abundant at higher levels.

One of the dominant genera at the Head, the clam *Macoma*, illustrates the difference between the two feeding types quite nicely. The ancestors of most clams were suspension feeders, but *Macoma* (see Coan 1971) belongs to a group which took to vacuum-cleaning the surface with its long intake tube (incurrent siphon), then filtering the food out of the water current with its gill. Of the two common species on the mudflat, one, the white sand clam, *Macoma secta*, retains this deposit-feeding habit; it takes up particles of sand from the bottom, and obtains its nutriment mainly from the bacteria which live on them. The other species, the bent-nosed clam *M. nasuta*, has largely but not exclusively reverted to a suspension-feeding mode of life. It mainly eats diatoms and other materials suspended in the water, but it has been observed feeding on the surface, evidently taking up benthic diatoms. That this difference does indeed exist, and constitutes a fundamental divergence between the two species, can be substantiated upon anatomical grounds: a host of structural adaptations directly related to the difference in feeding mechanisms characterize each species. Addicott (1952) found that at Elkhorn Slough *M. secta* was rare, a fact which we can now explain. *M. secta* at the Bodega Harbor mudflat is more abundant where the particle sizes are larger, hence more appropriate for its peculiar feeding mechanism, and Elkhorn Slough is a muddier habitat. *M. nasuta* becomes common in zone 3, but *M. secta* is not replaced by it. This makes sense, for although suspension feeding is not very profitable at higher levels, the deposit-feeding niche still exists lower down. One may note that neither species can feed when the tide is out, for both need a layer of water if their siphons are to take up any food. The deposit-feeding habit still explains why *M. secta* predominates at higher levels. It can fill its guts quite rapidly with sand from the immediate neighborhood. *M. nasuta*, on the other hand, must filter a large volume of water to obtain a full meal, and this takes time.

Macoma secta does not dominate the entirety of zone 2. Indeed, in many places it is replaced by yet another deposit feeder, one which likewise has a filter-feeding analogue lower down. Both are burrowing shrimps of the superfamily Thalassinidea: *Callianassa californiensis* and *Upogebia pugettensis*. Both *C. californica* and *M. secta* penetrate upward into zone 1, and indeed one might argue that the former species should not be considered a zone 2 form. *C. californica* lives farther up than does *M. secta*, and can feed without having to be covered by the tide. It obtains its food by sifting the sand through which it burrows, and this may account for its success at higher levels. *Callianassa* and other burrowing forms may be associated with a small clam, *Cryptomya*, which gets its food by filtering water from the burrow: since such burrows are common at several levels the clam quite naturally is widely distributed as well. It is not known precisely why the *Callianassa* and *Macoma* were not found in the same samples, but one might suspect that the burrowing makes conditions unfavorable for *Macoma*.

Other important animals in the mud of zone 2 and below are a wide variety of annelid worms. These are mainly tube-dwelling or burrowing animals which feed in various ways: some are active predators, some suspension feeders, some deposit feeders. Many annelids tend to be widely distributed, occurring in more than one zone. It should be evident that conditions within the mud do not change as abruptly as do the periods of exposure and submergence. It will require more extensive study before this group can be fitted in with our trophic analysis. The natural history of most local forms is poorly known.

ZONE 3

Here, the mode of obtaining food shifts toward suspension feeding. As we have mentioned, *Macoma nasuta* becomes more abundant. The other dominant organism in this zone is a worm of the phylum Phoronida, *Phoronopsis harmeri*. Particularly at the higher levels within the zone, it qualifies as a dominant both in a quantitative and qualitative sense. It dwells in a long, upright tube impregnated with sand grains, and obtains its food by sticking a crown of greenish tentacles into the open water at the surface. The dense clusters which it forms have been shown to help stabilize the mudflat, preventing the mud from washing away. They also provide something of a refuge for small animals that can live among the tubes (Hamner MS). One might expect the number and diversity of filter-feeding clams to increase at this level. Except for *Macoma*, in which this trend was indeed apparent, our sample was too small to be anything but suggestive. Part of the problem is that the clam populations have been decimated by people who dig them for food. Yet in spite of the fact that no definite zonation could be shown, the lower samples did show more species of clams.

In its lower part, we begin to find some of the animals characteristic of zone 4, such as the mudshrimp *Upogebia*. But the densities of such organisms are low, suggesting that albeit the boundaries are not abrupt, they are distinct.

ZONE 4

This zone (Fig. 5.5) displays a continuation of the filter-feeding economy, but one modified by increased influence of predators. The population densities for *Macoma* (both species) and *Phoronopsis* begin to decline at the mean lower low water level where zone 4 begins. The moon-snail, *Polinices lewisii*, drills holes in *Macoma* and other bivalves. The action of the tides may restrict its activity to lower levels, hence reducing the predation pressure above. Unfortunately, the exclusion experiments which would substantiate this conjecture have yet to be carried out. Sharks and rays and other fishes are important predators here too. A greater influence of predators should favor those animals best able to avoid them, and this may help to account for the presence of certain "deep burrowing" suspension feeders. *Upogebia*, with its burrow extending some distance below the level of the surface, is a good example. Another is the "innkeeper," *Urechis caupo* (Fig. 5.6), an echiuroid worm whose phallic appearance has rendered it a favorite topic for conversation among marine biologists. This creature digs a long, U-shaped burrow, through which it pumps water by peristaltic constrictions of its body. Near the entrance to the borrow it spins a fine

Fig. 5.5 Aspect of Zone 4 at near low tide.

net of mucus, with which it gathers food from the current. In their account of the natural history of *Urechis*, Fisher and MacGinitie (1928) describe as well the symbionts which commonly live with it. A small fish, *Clevelandia ios*, uses the tunnel mainly for shelter, and probably does little harm to its host. On the other hand, a scale-worm, *Harmothoe adventor*, and a pea-crab, *Scleroplax granulata*, in addition steal some food from the echiuroid. *Cryptomya*, the clam mentioned above as an associate of *Callianassa*, sticks its siphons into the burrow, and can, therefore, exist at a much greater depth than would otherwise be possible. A number of other mudflat forms (*Upogebia*, various annelid worms) have their own characteristic symbionts, substantially adding to the diversity of the system. A burrow of considerable depth, and one that is permanent, forms a very good home for such associations of animals, and it is hardly surprising that these associations have repeatedly evolved. They might be looked upon as the characteristic feature of the mudflat economy.

Protection and stability are very plausible reasons for the

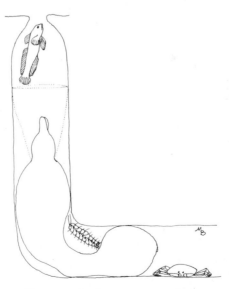

Fig. 5.6 Innkeeper worm in burrow with commensals.

development of symbiotic associations such as those we see in
the burrowing animals. The same basic phenomenon can develop above the surface as well. In zone 4 and somewhat
lower, especially in pools and shallow channels, one finds extensive beds of the eelgrass *Zostera*. This flowering plant is
important as food, whether eaten directly, as it is by certain
water-fowl, or as a source of detritus when it decays. It also
influences the community structure because the rhizomes help
stabilize the mud, and because both rhizomes and leaves provide refuge for small animals. During the summer especially,
when the eelgrass is growing rapidly, one can find a rich fauna
and flora living on and among its leaves. The blades support
a rich growth of diatoms and sessile animals. The motile animals living there tend to be green like their background, and
are hard to see unless one knows how to look for them.
Syngnathus, the pipe-fish, a relative of the sea horse, looks
very much like a blade of *Zostera* itself. A slug, or sea hare,
Phyllaplysia taylori (Fig. 5.7), crawls about on the leaf, eating diatoms along with some of the leaf itself. Its green almost
perfectly matches the leaf color, and its dark, thin lines
might be compared with leaf veins. One may also find various
crustaceans tinged with green and living in much the same
manner, such as a shrimp, *Spirontocaris*, an isopod, *Idothea
resecata*, and a skeleton shrimp, *Caprella californica*. Such resemblance to the substrate is hard to explain other than as camouflage. Associations with a plant and camouflaged symbiotic
animals have evolved elsewhere: the *Sargassum* community of
the Sargasso sea; organized about a floating seaweed with animals that closely resemble it, is a text book example. We don't

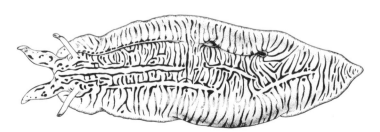

Fig. 5.7   Zostera slug. (*Courtesy of R. D. Beeman.*)

really know why such biotas have evolved in some places and not in others. Neither can we say for sure why the Bodega Harbor mudflat contains so many symbiotic associations both above and below the surface.

The mudflat in summer supports a fair crop of moderate-sized algae, such as *Ulva* and *Enteromorpha*, and a rich growth of diatoms. Near the *Zostera* bed, and extending to somewhat higher levels, one finds pools which remain more or less full when the tide goes out. These contain a good concentration of plants and small invertebrates, both herbivores and carnivores. The nudibranch mollusk *Hermissenda crassicornis*, a voracious predator, builds up large populations here. One can also find various slug-like animals related to both them and the sea hare that lives on *Zostera*. *Haminoea*, a green form with a thin external shell, feeds on plant materials; some years it is abundant, some years it has not been found here. The little *Aglaja*, which in the field looks like a large raisin, is a carnivore. Both *Haminoea* and *Aglaja* can burrow, but they feed on the surface. The distribution of these and other gastropods suggests a greater activity of surface-feeding animals at the lower levels.

Non-Resident Vertebrates and Insects

We have separated our account of birds and most fishes from that on the zones because these are transients who only feed there, rather than animals which reside in one zone or another. Since we are examining the community from the point of view of how food is obtained, this manner of organizing the materials seems justified. It should be obvious that one group, the fishes, will tend to feed when the tide is in, and that its importance will decrease at progressively higher levels, as a function of the time available for feeding. Conversely, the birds, even the diving ducks, will tend to have greater access to the higher level food items. Insects and a few mammals are basically terrestrial organisms which may range somewhat into the marsh and even the mudflat. Hence, we can regard these organisms, too, as members of the mud-flat community only in a partial and qualified sense.

We can provide only a little information on a few of the more abundant fishes (Hamner MS). The starry flounder, *Platichthyes stellatus*, feeds on the bottom, taking mainly polychaetous annelids but also a variety of mollusks, crusta-ceans and algae. Various species of embiotocid perch seem to be mainly omnivorous bottom-feeders. The bat ray *Mylio-batis californicus* eats polychaetes, crustaceans and bivalves; rays can feed on animals that live somewhat beneath the sur-face, and are important predators on clams. Gut analyses on the small leopard shark, *Triakis semifasciata*, revealed a ten-dency to feed on bottom-dwelling crustaceans. The shark stomachs also contained the deep-burrowing *Upogebia* and *Urechis*, along with their commensals, suggesting that this animal can feed by suction. In fact, biologists have found that a suction pump ("slurp-gun") is a very effective instrument for collecting *Upogebia*. We have recovered whole *Urechis* from the guts of other elasmo-branch fishes too, so the phe-nomenon may be quite common.

Many of the birds at Bodega are waders, seen feeding in shallow water or on the exposed surface of the mudflat. These waders range in size from egrets to small members of the

Fig. 5.8 Black-crowned night heron in cypress tree.

sandpiper family. The common egret, *Casmerodius albus*, and the snowy egret, *Leucophoyx thula*, occur on the mudflat most of the year. Both of these are pure white, but one can recognize the common egret by its larger size, and by its yellow, rather than black, bill. Both feed heavily on shrimp, fish, crabs and snails. The black-crowned night heron, *Nycticorax nycticorax* (Fig. 5.8) is squat and smaller; it feeds at night and can be observed in the late evening perched on the limbs of trees near the water. Like its relative the great blue heron, *Ardea herodias* (Fig. 5.9), it also is common in the freshwater marsh, and frogs, mice and crustaceans make up a great part of its diet.

The sandpiper family includes a number of common species: the marbled godwit, *Limosa fedoa*, willet, *Catoptrophorus semipalmatus*, dowitcher, *Limnodromus griseus*, and the tiny, least sandpiper, *Erolia ptilocnemus* (Fig. 5.10). All of these prey heavily on small invertebrates of the tidal flats,

Fig. 5.9   Great blue heron

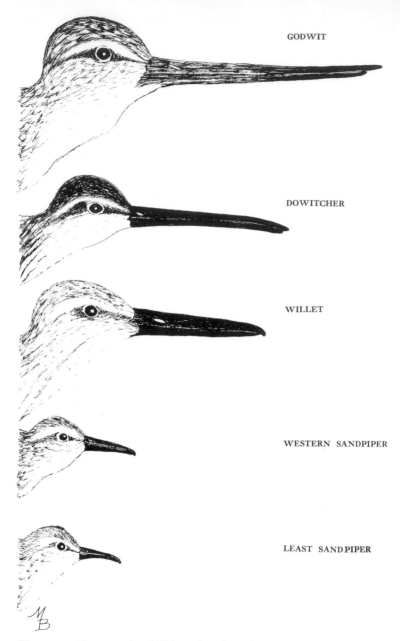

GODWIT

DOWITCHER

WILLET

WESTERN SANDPIPER

LEAST SANDPIPER

Fig. 5.10   Comparative bill lengths of sandpipers; *from top to bottom:* GODWIT, DOWITCHER, WILLET, WESTERN SANDPIPER, LEAST SANDPIPER.

but their differences in size and in the shape of the bill reveal differences in feeding habits which allow them to specialize and reduce competition between the species. The marbled godwit's bill is long, slender, and slightly upturned at the end, well-adapted to probing deeply in the mud. The dowitcher has a shorter bill, and the willet an even shorter one, but still considerably longer than the bills of birds which pursue insects on land. The least sandpiper takes animals at or near the surface, and can be seen running swiftly in and about the feet of other shore birds.

The structural differences provide indirect evidence that the shore birds have diversified and specialized in their manner of feeding. Recher (1966) was able to substantiate this by gut analyses and field observations on birds on mudflats at Bodega Harbor and elsewhere in northern California. He estimated that the least sandpiper was able to penetrate less than 1 inch into the mud, the willet 2 inches, the dowitcher 2 to 3 inches and the marbled godwit 4 to 6 inches. Different kinds of animals live at different depths in the mud, and the feeding level would affect the diet. Recher's gut analyses did in fact show that the birds were using different kinds of food. For instance, amphipods (small crustaceans) made up 21% of the gizzard contents of the least sandpiper, 49% for the willet, 78% for the dowitcher, and 82% for the marbled godwit. Recher also showed that seasonal migration patterns result in species with closely similar feeding habits occupying "the same niche" but at different times. The least sandpiper and the western sandpiper, which migrate at somewhat different seasons, may both be observed at Bodega and at Palo Alto, but not together. The least sandpiper is present in summer and fall, the western in winter. They resemble one another closely and have nearly identical food habits. Thus we have another example—like the limpets of rocky shores—of an assemblage of organisms which appear at first sight to be occupying the same niche, in apparent contradiction of the competitive exclusion principle. Yet the competing forms actually make their living in a somewhat different way. American avocets, *Recurvirostra americana* (Fig. 5.11), have recurved bills like those of godwits, but are not considered close relatives of godwits or other members of the sandpiper family. Animals living in more or less the same circumstances need not diverge in all respects.

Most of the birds that feed on the substrate fauna of the mudflats are migratory, and utilize this rich food source only temporarily. Yet they are vitally dependent upon it. Without these feeding areas the total number of shore birds would surely decrease. As more and more mudflats are dredged out to become boat harbors and marinas, we can expect the number of these birds to decline. Those who derive pleasure from watching these beautiful animals can enjoy an innocuous but

Fig. 5.11   American avocet

inexpensive form of recreation; they are hard pressed to defend their interests from those persons seeking to "develop"
the local recreation industry.

In deeper water, there occur concentrations of a group
of birds known as swimming divers. They are usually seen
swimming in leisurely fashion on the water's surface, or just
sitting there. All obtain food by diving under the water. Many
are only seasonal residents. Birds in this group include three
species of grebe: the western, *Aechmophorus occidentalis*
(see Fig. 7.3), the eared, *Podiceps caspicus,* and the horned
grebe, *P. auritus,* all common in winter and spring. The western grebe is the largest and the most common. Grebes are
sometimes called "helldivers" because they can stay under
water for a long time; they feed on bottom invertebrates.

Common loons, *Gavia immer,* and arctic loons, *G. adamsii,* are fish-eaters. They are also winter residents; those that
stay all summer do not breed at Bodega. Many native Indian
tribes attributed supernatural powers to them and considered
them reincarnated ancestors. They have a very chilling cry.

A number of ducks are swimming divers. The white-
winged scoter, *Melanitta deglandi,* and red-head ducks, *Aythya americana,* are seen all year. But the surf scoter, *Melanitta perspicillata,* golden eye, *Bucephala clangula,* ruddy
duck, *Oxyura jamaicensis,* and lesser scaup, *Aythya affinis,*
are winter residents. The latter three divide their diet between plant and animal matter, the others eat primarily bottom invertebrates. Also present are large numbers of American coots, *Fulica americana;* superficially these look like
ducks, but have no webbing between their toes and are not
related to true ducks.

Another group of birds conspicuous around the harbor
are the "high divers." These birds cruise above the water until
prey (usually fish) is sighted, then fold their wings and dive
into the water after it. This manner of feeding has evolved

independently in many taxonomic groups of birds, and includes a fair range of body size. Among the smaller high divers are the trim-looking terns: Forster's tern, *Sterna forsteri*, is common in late summer, its larger relative the Caspian tern, *Hydroproghia caspia*, in fall and winter. Belted kingfishers, *Megaceryle alcyon*, are common throughout the year, and they breed at the harbor. They hunt over the harbor much as sparrow hawks hunt over grassland, hovering fifteen to twenty feet above the surface, then plunging dramatically down to it—coming up with a fish instead of a mouse. These beautiful and fascinating birds are found not only on coasts but on inland rivers, lakes and ponds—any place where they can exercise their remarkable skill at fishing. George Salt and Daniel Willard (1971) studied the hunting behavior of Forster's terns on tidal ponds not far from Bodega. They found that the hunting efficiency varied greatly with several different factors. For instance, the capture success (number of dives/captures) on a clear day in April or May was only 10.7%, but on a clear day in October or November it was 49.3%. They found, though, that as capture success increased, the total amount of reward for the tern was less because the size of the fish caught declined. In April and May when success was low, Salt estimated that each tern was garnering 1.34 weight units (arbitrary) every second of hunting time but in November and October they got only 0.12 weight units/sec. As the prey got smaller, due to seasonal changes in population structure, the terns became better at catching it. Even so, their attack efficiency was not sufficient to counteract an overall decline in food getting. Larger high divers include the brown pelican and osprey. Both brown and white pelicans are at the harbor from late summer through winter, but only the brown, *Pelecanus occidentalis*, is a high diver; the white, *P. erythrorhynchos*, is a swimming diver. The osprey, *Pandion haliaetus*, is the most majestic of the diving birds at

Bodega. Its numbers have been dwindling in North America for the last twenty years. One possible cause for its decline is discussed in Chapter 7.

Several gull species are opportunistic omnivores, and are common in many Bodega habitats besides the harbor. The western gull, *Larus occidentalis,* is a year-round resident; the ringed-billed gull, *L. delawarensis,* and the similar-looking California gull, *L. californicus,* are abundant only in fall and winter. They take both living and dead materials: fish, invertebrates, plant materials and garbage. The smallest of the gulls seen at Bodega is the Bonaparte's gull, *Larus philadelphia.* All the other species of gull, and especially the young, are very similar to each other in appearance and difficult even for the experienced bird-watcher to separate. The Bonaparte's gull, on the other hand, is easily distinguished; it is smaller and has a distinctive black head. It is the only blackheaded gull commonly seen on the West Coast. Large flocks of these birds can be seen at Bodega in late winter and spring.

The water bird fauna at Bodega changes continually with the season. Only a fraction of the total bird species reside there throughout the year. Many spend only a few days on their way to a seasonal home either north or south. Others stay for an entire season, making Bodega their summer or winter home at one end of their migratory path. Many of the winter residents nest in the Arctic during the summer. Two or three times as many species are seen in December as in June. Seasonal residents, although only temporary, are not external to the Bodega Head ecosystem. Nor does their brief period of residence mean that they are unimportant. Indeed, many year-round residents, such as aestivating snails and seasonally-abundant insects, restrict their activities there to a fairly brief period.

In the marsh one finds something of a mixed bird fauna. Many of the wading birds of the tidal flats feed there on in-

sects and marine invertebrates such as crabs. The most common terrestrial bird that feeds in the marsh is the barn swallow, *Hirundo rustica*, which winters in South America but
is abundant in the summer over the grassland, and near buildings (where it nests), as well as in the marsh. In an area as
small as four to five acres, as many as forty barn swallows
may individually swoop and dive in the cloud of insects over
the marsh at one time. When the tide is in, some of the wading birds such as godwits and sandpipers leave the marsh; but
others, such as the willet, feed on invertebrates on the pickleweed above the waterline.

The familiar small mammals of grassland—meadow vole,
deer mouse and shrew—are also present in the marsh. Population densities are far higher here than in grassland. Nights
when the tide is not high are periods of peak activity for
mammals: then the low, wet part is dominated by shrews,
the middle by voles, and the inland part by deer mice. Yet
none of these animals actually burrow or nest in the marsh.
They all retreat during rest periods to tunnels in very high
ground behind the marsh. They seem to roam at random
within their zones, and no well-marked runs have been detected. Of the ninety-five animals trapped during successive
nights, not one was captured twice. The harvest mouse,
Reithrodontomys megalotis, is a salt marsh inhabitant elsewhere along the Pacific coast; but in our study area this species was found in the fresh-water marsh and rarely in the
grassland and did not occur in the salt-water marsh. Small
mammals in the salt marsh are preyed upon by the great blue
heron, an occasional marsh hawk, owls, and grassland raccoons. Both the heron and the raccoon feed on fish, crabs and
other small animals as well. Raccoons have made so many
trips to the marsh that they have worn a trail through it.
Their tracks may be found on the mudflat as well.

Insects are very abundant in the marsh during summer.

We have very little data on insect populations at Bodega, but a study by Davis and Gray (1966) of a North Carolina salt marsh showed that insect populations were small until late April, then climbed rapidly to reach a peak in June. Population sizes remained high until fall, when the marsh became dormant; few insect species reached population peaks in the fall. When the marsh was exposed at low tide, Davis and Gray found distinct zonal distributions not only of species, but of entire orders. The order Homoptera (leaf-hoppers and allies), for example, dominated the insect fauna (90% of all those caught) in the lowest part of the marsh and declined in abundance inland. The Diptera (true flies) exhibited the converse pattern, increasing in abundance inland until they dominated the driest part of the marsh (44% of those caught). But when the water inundated the marsh, the insects left and the zones could no longer be observed. Davis and Gray experimented on ten species of marsh insects by placing clumps of marsh vegetation in a specially constructed chamber, introducing an insect, and slowly adding water from the bottom to imitate a rising tide. Only one species could tolerate as much as four hours of immersion if trapped; the others drowned. All displayed one behavior pattern or another which kept them out of water. They would climb plant stems, hop, swim, or walk on the water surface to higher, drier vegetation. Marsh insects are important components of the marsh association, for they feed upon quite a variety of plant and animal materials, and are themselves fed upon by other animals. But, like the birds, they visit the marsh when the conditions are right for feeding, and spend much of their time elsewhere.

Community Structure

This is not the place to review the theory or philosophy of the community concept. We would like, however, to place the materials covered here in perspective, and see if we can add a little to the general remarks we provided in the introductory chapter. Just what *is* a community? This turns out not to be a single question, but two: "community" here may mean either a *particular* community (such as the mudflat community in Bodega Harbor), or it may mean a *kind* of community (the mudflat community wherever it occurs). Some workers view the community (or any given community) as fundamentally a statistical entity. Thus, each community would be an assemblage of organisms which can be characterized by the relative abundance of certain species. Others take more of a functional, hence qualitative, view of things. A community would be a group of organisms living together in a particular way, and we would classify communities together into groups, and draw lines between them, according to how they are organized and not according to what their components may be.

Either way of viewing the matter may be useful, depending on one's goals. For analysis, the former may be preferred, but let us see what we can do with an attempt at synthesis along functional lines. Viewing communities in terms of how they are organized allows us to treat them in economic and sociological terms. At the risk of some anthropomorphism, let us see how we can treat a natural community much as we might a "community" in the more familiar sense—an association of human beings and their resources. This is more than a mere far-fetched analogy: many ecological principles apply to all organisms, and men are, beyond dispute, organisms. Consider the little town of Bodega Bay (population something on the order of 500, if one doesn't take the census figures too seriously). Bodega Bay resembles other human communities all up and down the coast, in spite of variations. Its structure is determined largely by economic forces, and, not surprisingly, one of the most conspicuous aspects of its structure is zonation. Along the water we find a fishing industry—dominant producers—and by the highway another very significant source of energy, various tourist services. Land for either of these endeavors being scarce, it fetches so high a price that few can afford to live right by the highway or on the water. Hence, private residences and facilities like the church and school tend to occupy a slightly higher, or more peripheral, position. And a clear distinction can be drawn between the places occupied by year-round residents (who need inexpensive quarters) and visitors (who want a view). Beyond the immediate community proper, we find a sparsely populated and economically monotonous region, given over mainly to agriculture.

The foregoing analogy allows us to dismiss, as rather beside the point, a number of philosophical controversies about the community. We don't say that the town of Bodega Bay is not "real" because we aren't sure whether to consider the Bodega Marine Laboratory a part of it. And we have the best of reasons for classifying the town as a "fishing village," namely, that it tells us that a very important activity goes on there. It doesn't matter that elsewhere along the coast other kinds of seafood are taken, nor does it make sense to object that the mix of agriculture and tourism varies from place to place. And it should not bother us that, for some purposes but not for others, we may wish to recognize subunits within a given local economic system.

Turning to the natural assemblage within the harbor itself, we could provide it with various descriptive labels. We might simply refer to it as a "mudflat community," reflecting the physical conditions which do indeed determine much of what goes on. A good alternative would be a "*Macoma* community," after a numerically and functionally dominant organism. For other purposes an economic name might be preferred: a "suspension- and deposit-feeding community," after the main "industry." We might wish to view the mudflat as an assemblage of several different communities, such as a "*Zostera* community." We see, too, that the mudflat does not exist in isolation: it exchanges resources with other places, and its inhabitants include both full-time residents and organisms which spend some of their time elsewhere.

An economic perspective may help us to view the community in dynamic terms rather than as just a static collection of individuals. For instance, we might ask what determines the zonation of the organisms within the harbor. On the basis of studies not unlike our own, carried out in Tomales Bay, Johnson (1970, 1971, 1972) found that at increasingly lower elevations, new kinds of organisms appeared. But, to a considerable extent, the high level forms did not drop out. Hence, he could document a species diversity gradient; and what we have seen of both rocky and muddy intertidal situations at Bodega Head fully backs him up. Johnson sought to explain his data in terms of a kind of additive successional model, based in part on the work of some other ecologists. He reasoned that at higher levels the evironment is less predictable, and that conditions there are less stable. Hence, the higher level assemblages should be more frequently disturbed or wiped out by the action of the physical environment. Now, if one group of organisms can only live where certain others are present, we might find a kind of orderly development, or succession, taking place, one in which more species are gradually added, in turn making the existence of yet others possible. The diversity would be a function of the time since the habitat had been disturbed, and the length of this period would increase with depth. Hence, the lower the zone, the more species.

Johnson's hypothesis does seem quite plausible when we consider some of the communities within the Bodega Harbor system. For one thing, his theory doesn't take into account the interdependence of organisms. If a part of the mudflat should become cleared of *Zostera*, then the numerous species associated with it should be absent until it returns. Extension of the same basic line of reasoning would be easy for quite a number of other associations.

And when we consider zonation in the light of our trophic analysis, another hypothesis suggests itself (Ghiselin MS). The suspension feeders get "subtracted" at higher levels, not because they are unable to withstand the instability, but rather because the tide is in so briefly that their way of obtaining food becomes unprofitable. But deposit feeders prosper at both higher and lower levels, because their mode of feeding requires less time under water. Competition and predation would together account for a gradual decrease in the abundance of deposit feeders lower down. High up, in the marsh, even the deposit feeders drop out, and the remaining source of food, namely, the surface algae and other plant materials, becomes the basis for most of the economy. It is not simply the unstable environment that causes there to be so few kinds of snails and other animals in the marsh or on the higher intertidal rocks, but a lack of opportunity for organisms which make their living in a different way. Just as with increasing elevation on the mudflat, we see fewer and fewer niches being occupied, so as we climb upward from the water in Bodega Bay, we reach a point where agriculture becomes the sole profession. The extent to which the parallelism between the human and the natural economy holds true can of course be debated. But once again we see how a correlation may be reconciled with more than one plausible hypothesis.

6

The Fresh-water Marsh and Related Habitats

Of Pyramids and Limiting Factors

This chapter discusses several habitats: the large fresh-water marsh on the east side of the peninsula, the brackish ponds near the harbor, a fresh-water pond in the dunes, seasonally wet depressions inland, shaded gulleys, and seeps.

Water enters the fresh-water marsh at its northwestern end from seeps and springs. It meanders in open rivulets toward the southeast at a speed of ¼ mph, and eventually empties into the harbor. This water is fresh (always below 0.05% salinity), cold (45°–65°F), and acidic (ph 5.5–6.0 in September). The water saturates the soil for many meters away from the streams, but does not appear on the surface (though vegetation is so thick, it is difficult to see the soil surface). As distance from the streams increases, the soil becomes drier. Plant species replace one another in zones at different distances from open water, but because the rivulets are so many and take winding routes, the zones meet and twist about each other in complicated patterns. The marsh covers about twenty-five acres; comparison of aerial photographs taken in 1953 and 1969 shows that its extent has changed very little, if at all, over sixteen years.

Several brackish ponds, isolated from the harbor by a road, exhibit a different flora and fauna than either the harbor or the fresh-water marsh. A fresh-water pond in the dunes was created when construction workers removed fill for laboratory construction in 1966. They removed enough sand to reach the water table, and now a pond covers about an acre.

Seasonal ponds or wet areas form in depressions in the grassland and dunes and in roadside ditches. The soil in such depressions remains wet longer into summer than does the surrounding soil, but it eventually dries; this seasonal lack of abundant water prevents many fresh-water marsh species from establishing themselves, and other species take their place.

On the steep, shaded banks of gulleys, shrubs become more abundant and the vegetation is very dense—in imitation of the northern coastal scrub which dominates the coast further north (see Chapter 2).

Seeps, wherever they occur, permit strange plant associations to develop; associations in which members of ordinarily quite different habitats mingle and grow side by side.

Ponds and springs have been used as test ecosystems by biologists to determine, in detail, how food chains really operate. These ecosystems proved ideal for such research, because they are simple and not too diverse to prevent detailed work on each species present: what it ate, how much weight it gained, how much oxygen it used up, and what ate it, and how often. And, despite the relative simplicity of the ecosystems, an enormous amount of other data could be collected to round out the total picture.

One lesson that stands out from such studies (for example, those by Lindeman 1942; Odum 1957; and Tilly 1968) is that energy passes out of the system all along the food chain. As mentioned in Chapter 2, much of the chemical energy trapped by plants is used by them to grow and maintain themselves, and only a fraction of it is passed on to herbivores which graze on them; only a fraction of the herbivores are preyed on by carnivores; and so forth. A general rule of thumb is that energy is reduced in magnitude by 100-fold in the single step from producers to herbivores, and 10-fold for each step thereafter. In addition, as we have already seen for the Bodega grassland (Chapter 2), plants are able to convert only about 1% of radiant energy into chemical energy. Thus, if we consider only a very simple food chain of three levels—plant, herbivore, and carnivore—the amount of energy available for the carnivore to use is only 1/100,000 of radiant solar energy:

$$\begin{array}{l}\text{energy} \\ \text{available to} = \\ \text{carnivore}\end{array} \begin{array}{c}\text{energy passed,} \\ \text{sun to plants}\end{array} \times \begin{array}{c}\text{energy passed,} \\ \text{plants to herbivores}\end{array} \times \begin{array}{c}\text{energy passed,} \\ \text{herbivore to} \\ \text{carnivore}\end{array}$$

$$= (1/100) \times (1/100) \times (1/10)$$
$$= 1/100,000$$

There are several reasons for the apparently inefficient use of the available energy. First, if energy is to be passed on, it must be fixed or trapped in living tissue. But a large portion of the energy is used merely in keeping that tissue alive; it neither goes toward the production of new tissue, nor is it consumed by other organisms, so it does not appear in our efficiency equation. Second, energy transfers in biological systems result in as much as 50% of the available energy being lost as heat. Third, the efficiency with which consumed food is assimilated is much less than 100%; that is, a certain portion is passed through the gut relatively unaltered as feces, and this portion falls to the decomposer food chain. Herbivores utilize food inefficiently, at least in comparison to the organisms of higher trophic levels.

Odum has shown that the weight of organisms at each food chain level declines abruptly. Since weight is roughly proportional to caloric energy, the biomass pyramid in Fig. 6.1 is equivalent to an energy pyramid. More complex food chains than the plant-herbivore-carnivore one discussed above include secondary or tertiary carnivores (carnivores that eat carnivores). The fraction of energy available to those carnivores would be 1/10 or 1/100 of the above. Clearly, it takes a lot of plant matter to support relatively few herbivores and

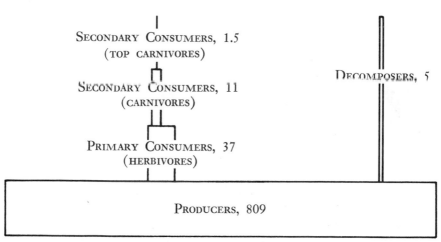

Fig. 6.1 Biomass pyramid for Silver Springs, Florida. Figures are grams per square meter, dry weight. Redrawn from Odum 1957.

even fewer carnivores. This tremendous reduction of energy from level to level may well account for the fact that food chains rarely have as many as five levels.

Biological investigations of other lake, stream, or pond ecosystems have sometimes revealed that efficiency of plant productivity was unexpectedly low. The reason for low productivity often turned out to be low levels of certain critical nutrients, such as nitrogen, phosphorus, or molybdenum. If additional amounts of these limiting nutrients can be added, plant productivity, and ultimately fish population size and growth, increases sharply. When the limnologist Charles Goldman (1960), for example, added only enough molybdenum to Castle Lake in northern California to raise its concentration by 100 ppb (that's parts per *billion*), the rate of plant productivity within it tripled in four days' time.

What the nutrient status of the Bodega fresh-water marsh is, we don't know at this time; nor do we know the details of the food chain. However, the next section does present the general interactions going on in the marsh.

The Fresh-water Marsh and 300 Million Years

To document zonation of plants in the fresh-water marsh, we sampled the vegetation along a transect which stretched nearly 200 m from the outer, dry part of the marsh in to the center, with open water. The presence of plant species was noted in every 5 m segment of its length. The results appear in Fig. 6.2.

The outermost, driest zone is a broad one, dominated by three species (Fig. 6.2) almost to the exclusion of any others: velvet grass, *Holcus lanatus* (Fig. 6.3), rush, *Juncus leseurii*, and cinquefoil, *Potentilla egedii* (Fig. 4.13). The latter two species also occur in wet depressions in the dunes, but in the

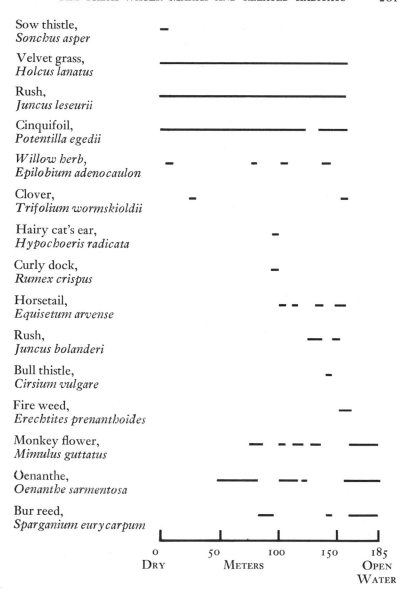

Fig. 6.2 Presence of plant species along a 185 meter transect from the outside of the fresh-water marsh, on dry land, in to the center, with open water.

dry zone they reach their most luxuriant growth. Clumps of rush, several feet in diameter at the base and as much as three feet in height, are so closely packed that walking is difficult. Intertwined with rush stems near the ground are prostrate shoots of cinquefoil, whose leaves are dark green on one side, silvery white on the other. Velvet grass plants are common, but do not form pure swards as they do in grassland depressions.

Several species appear as scattered individuals throughout this rush community, but they are not characteristic members of it and all can be found elsewhere. Sow thistle, *Sonchus asper*, and clover, for example, are more common in grassland; willow herb, *Epilobium adenocaulon*—which sheds hairy seeds that float through the air like those of cottonwood —hairy cat's ear, *Hypochoeris radicata*, and tall curly dock, *Rumex crispus*, are more common along roads and paths.

Nearer to open water, several new species become common: the white-flowered, succulent-stemmed *Oenanthe sarmentosa*; monkey flower, *Mimulus guttatus*; another rush, *Juncus bolanderi*, with clusters of brown flowers much denser than those of *Juncus leseurii*; fire weed, *Erechtites prenan-*

Fig. 6.3   Velvet grass

thoides, with tall stems and narrow, serrated leaves; and small horsetails, *Equisetum arvense*.

These small, herbaceous horsetails are the sole evolutionary remains of a large group of plants that dominated the world's forests about 300 million years ago. The tree species were as tall as thirty feet, with trunks over a foot in diameter. They became extinct 200 million years ago, possibly as a result of a change in world climate from tropical, warm and wet to temperate, cool and drier. Conifers dominated world forests after that, until further cooling and drying, when they retreated to higher latitudes. The temperate forests then became dominated by a group of plants whose ancestry seems to go back less than 150 million years—just a squiggle on the graph of geologic time. In that short amount of time this group of plants, the flowering plants, came to dominate not only the temperate and tropical forests, but grasslands, marshes, deserts, and alpine zones as well. It isn't known why one group adapted and moved into every major habitat: wet, cold, dry, hot, fresh, saline, shaded, exposed. Conversely, we cannot say why most horsetails failed to evolve the properties necessary to maintain themselves.

Finally, at the very edge of open water, or actually in it, is a community completely different from that dominated by

Fig. 6.4 Bur reed

rush. This community is dominated by three large grass allies, all difficult to tell apart by leaves alone: umbrella sedge, *Cyperus eragrostis*, bulrush, *Scirpus microcarpus;* and bur reed, *Sparganium eurycarpum* (Fig. 6.4). Small, succulent, weak-stemmed herbs include speedwell, *Veronica americana*, with rounded, sessile leaves; brass buttons, *Cotula coronipifolia;* dark green water cress, *Rorippa nasturtium-aquaticum;* and the marsh pennywort, *Hydrocotyle ranunculoides.*

Moving water is clear and algal growth is minimal, but in pools isolated from the rivulets, the water is crowded with algae and with aquatic flowering plants. Soft, filmy strands of *Spirogyra* and tougher, floating clumps of *Oedogonium* are the common green algal genera; a number of unicellular and filamentous diatoms are visible only with a microscope. Shoots of *Oenanthe* may penetrate the surface, and hundreds of individuals of duckweed (*Lemna valdiviana*, Fig. 6.5)—one of the smallest flowering plants in the world, about 2.5 mm long and wide—float on the surface. In large ponds, such as the acre pond in the dunes, two species of larger aquatic flowering plant, pondweed or *Potamogeton*, are common.

If the pond water is more brackish, due to proximity of the harbor, different terrestrial and aquatic species are present. One of the most striking is the green alga *Enteromorpha*, whose tough, floating clumps may nearly cover the water surface in spring and summer. A common bird around the edges of these pools is the killdeer, *Charandrius vociferus*. The

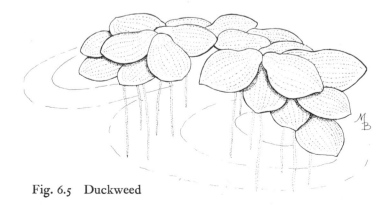

Fig. 6.5   Duckweed

female builds a ground nest near the water edge, and when disturbed by man or predator she will leave the nest and chicks, drop one wing in an imitation of helplessness, and cry out. She continues the act while hopping away from the nest, drawing the predator with her. If he gets too close, she stops feigning and flies off; meanwhile, her young have been protected. Killdeer are also commonly seen on tidal flats and in agricultural areas inland.

The lanky, yet graceful, great blue heron mentioned in Chapter 5 is commonly seen in the fresh-water marsh. Groups of fifteen to thirty can be seen feeding in the deeper areas of the marsh among tall sedges and reeds. They are reported, in the literature, to feed on mice, amphibians, and fish, and likely do so here in this marsh. They roost in upper branches of a grove of nearby Monterey Cypress planted at the north end of the marsh.

During fall and winter, several species of black-colored birds congregate in enormous flocks to roost in dense parts of the marsh. We have estimated that more than 100,000 birds roost in the marsh at that time. They arrive at dusk in groups ranging from five birds to several hundred; at dawn their chirping reaches a crescendo and they take off in a gigantic, dark, undulating train—an awesome and spine-chilling spectacle, reminding one that this is the setting of Alfred Hitchcock's horror film, *The Birds*. The flocking habit of several bird species has been given a great deal of research attention lately. The adaptive significance of flocking is an interesting and partly unanswered question as yet.

Brewer's blackbirds, *Euphagus cyanocephalus*, make up the majority of this flock. They feed primarily on insects, spiders, and seeds; when feeding, the flock breaks up into smaller groups. Many can also be seen on rocks along the harbor, and at low tide it is not unusual to observe them in the mudflats. They are not often seen in the undisturbed grassland on the Head, but are common in inland pastures where they feed on fresh dung and grain.

The most commonly seen bird in the marsh is the long-billed marsh wren, *Telmatodytes palustris*. This active little bird hops among the bur reeds, emitting a constant chatter. It is primarily insectivorous, and is a year-round resident. Marsh hawks and sparrow hawks are frequently observed here too. The common gallinule, *Gallinula chloropus*, a close relative of the American coot, is a permanent resident of the marsh. Like the coot, it is omnivorous, but feeds heavily on insects such as aquatic beetles and bugs.

The small mammal species show quite a different pattern of abundance from that which we considered for the grassland, dunes, and salt marsh. The deer mouse has never been trapped or seen in the fresh-water marsh. Shrews and voles are regularly found only at the drier edge of the marsh. Apparently the most abundant small mammal is the harvest mouse, *Reithrodontomys megalotis*—a species found nowhere else on Bodega Head (except rarely in grassland). The harvest mouse is not found in drier areas with shrews and voles, but in damp habitats shaded by high, dense clumps of rush, where a layer of dead herbage four to six inches thick covers the ground. The traps we set on the ground, below the mat of herbage, never caught anything; only those set on top of the mat caught the mouse. The density of vegetation may have reduced the animals' likelihood of encountering a trap, for trapping was very poor (never more than 5% success, and once not one animal was caught for one hundred traps set).

Raccoons feed on amphibians and (probably) snails in the marsh; their tracks and scats are frequently seen. Musk-rats, *Ondatra zibethica*, are sometimes seen in the wetter areas of the marsh. Like its close relative the vole, the muskrat eats vegetative parts of plants rather than seeds; it seems especially partial to rushes and cattails. Muskrats grow to the size of a small house cat and are likely prey for the gray fox; their large size protects them from avian predators. They are excellent swimmers, an activity in which they are aided by partially webbed feet and a laterally-flattened tail.

The lined snail (see Chapter 2) is abundant in the freshwater marsh where, unlike populations in the grassland, it does not aestivate, but rather remains active and is even particularly abundant in summer.

In fall and winter, salamanders are reported by others to emerge and become an active part of the marsh community both as predators and prey; however, we have not yet confirmed this.

Seasonally Wet Depressions and the Ecotype Concept

Many of the species found in seasonally wet depressions are also found in the fresh-water marsh: several species of rush (*Juncus*), cinquefoil, velvet grass, and brass buttons. A few fresh-water marsh species are absent from the depressions: bur reed and speedwell come to mind. Cattail, *Typha domingensis*, is one of the largest plants in seasonally wet areas, but it is rarely found in the marsh. This plant, like other members of its genus, can tolerate extremes in soil moisture and as a result is widely distributed in a number of dissimilar habitats. *Typha latifolia*, for instance, grows equally well along the cool, wet California coast and in the hot, dry central valley; it is found in areas wet with fresh water or those wet with brackish water. Other Bodega species also appear in diverse habitats: velvet grass, *Holcus lanatus*, grows in the fresh-water marsh, in seasonally wet depressions in the grassland, and occasionally in mildly brackish salt marshes; brass buttons grow in the fresh-water marsh; sea fig grows at the lip of grassland bluffs, in the splash zone of the intertidal, and in the dunes; rabbit foot grass, *Polypogon monspeliensis*, grows along dry roadsides and in seeps near the base of grassland cliffs.

How is it possible that one species can exist in such extremely different habitats? Perhaps, like man, any individual can adapt to a wide variety of environments. Another possibility (and the two are not mutually exclusive) may be that these widely distributed species are not genetically uniform; that the population of species X which exists in a salt marsh has a different hereditary makeup from that of another population of species X which grows in a fresh-water marsh. The two populations of X may consist of individuals which look very much alike—even identical—and the two may be capable of interbreeding to produce viable seeds, but differ both genetically and physiologically. The two populations have diverged, possibly in a short amount of time, so that tolerance limits to salinity are different. These two hypothetical populations are included in the same species, but each consists of variants within that species which are ecologically and genetically adapted to a particular environment.

Plant ecologists have a word for this sort of variant. The term ecotype was coined by the Swedish botanist Göte Turesson (1922). He collected seeds of species which had a wide range of habitats: from lowland, southern and central Europe, to northeastern Russia in the Ural Mountains. When he germinated the seed and grew the plants in a uniform garden in Akarp, Sweden, he found that members of widespread species were not uniform. Those plants which grew from seeds collected in warm, lowland sites often were taller and flowered later in the year than those which grew from seeds collected in cold, northern sites. Yet these variants had very similar flower morphology and were interfertile, indicating that they were all part of one species. Turesson called these variants ecotypes. Evidence accumulated since then indicates that most widespread terrestrial species have several ecotypes, each differing slightly in distribution, morphology, or physiology.

Coastal and inland ecotypes of the cattail, *Typha latifolia*, have recently been discovered by S. J. McNaughton (1966), a botanist now at Syracuse University. He collected rhizomes near Point Reyes on the coast and near Red Bluff at the northern end of the central California valley. In nature these plants are exposed to very different temperatures: on Point Reyes average summer maximum daily temperature is 60°F and leaf temperature might then be 85°F; near Red Bluff daily maximum summer temperatures average near 100°F, and leaf temperature might reach 120°F.

The differences between the ecotypes may have something to do with the chemical stability of protein molecules. Leaves are important plant organs; the vital metabolic processes of photosynthesis and respiration go on within their tissues, and these require the presence of many specific enzymes. Enzymes, like other proteins, are sensitive to heat; if temperatures become high enough they lose their activity and the organism suffers as a result. McNaughton wondered if the enzymes of the two populations showed different temperature tolerance limits. He grew both plants in a uniform growth chamber for several months, then tested the activity of several enzymes at 120°F. He did this by grinding up leaf tissue, incubating it at 120°F for various periods of time, then measuring the ability of the enzymes to attack a known amount of substrate.

He checked the activity of three enzymes. Two of them showed the same tolerance of temperature for the coastal and inland populations. The third, malate dehydrogenase, was very stable at 120°F from Red Bluff plants, but was quickly denatured at that temperature from Point Reyes plants. After thirty minutes' incubation at 120°F, the enzyme from Red Bluff plants lost only 10% of its activity, but the enzyme from Point Reyes plants lost 70% of its activity.

McNaughton has also shown that ecotypes of *Typha latifolia* growing at 6000 feet elevation in Wyoming, which experience a very short growing season, are more efficient in photosynthesis than plants from Point Reyes, where the growing season is essentially twelve months long. The efficiency was twice that of Point Reyes plants, and the basis for the difference was enzymatic.

Differences in enzyme tolerance limits may possibly explain or at least help us to understand other plant distributions at Bodega Head.

Shaded Banks and a Travelling Eucalyptus Story

The microenvironment of protected, wet hillsides and gulley bottoms differs significantly from the general Bodega Head macroenvironment. The coolness and wetness permit growth of some plants that grow far to the north or of some usually associated with densely shaded forest floors. Pacific reed grass, *Calamagrostis nutkaensis*, for instance, covers some of the gulley banks in large clumps; Bodega is near its southern limit, and it extends up the coast all the way to Alaska. Other large clumps on the hillsides are of the rust-colored sword fern, *Polystichum munitum* (Fig. 6.6), and sword ferns are commonly seen in damp, dark redwood forests along the coast of northern California. As already discussed in Chapter 2, several other species of these wet hillsides become more abundant and widespread further north and form a vegetation type Munz and Keck call northern coastal scrub.

Common shrubs include two willow species, *Salix*; wax myrtle, *Myrica californica*; buckthorn, *Rhamnus californica*

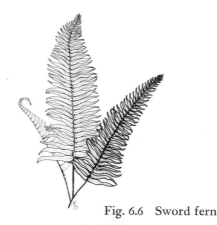

Fig. 6.6 Sword fern

ssp. *tomentella;* the introduced Himalaya berry, *Rubus pro-cerus,* which forms tall prickly thickets along shaded stream-banks; and the small shrubby monkey flower, *Mimulus au-rantiacus.*

Vines and canes include the viciously armed eglantine, *Rosa eglanteria,* and several introduced blackberry species.

The most striking herbaceous perennial is cow parsnip, *Heracleum lanatum* (Fig. 6.7), whose flowering shoots are five feet tall and basal leaves one foot across. Tall, purple-splotched stems of poison hemlock, *Conium maculatum,* the same plant that Socrates made famous, are occasional. Smaller, more colorful herbs include paintbrush with red-orange flow-ers, pearly everlasting with glistening clusters of white flow-ers, and a four-foot-tall vetch, *Vicia gigantea.* Live forever, poison oak, and bull thistle—all inhabitants of grassland—are common on these hillsides. Some fresh-watch marsh plants are found in the gulley bottoms: horsetail, water cress, chick-weed, and a number of delicate ferns. The rarest plant on the Head, the herb *Fritillaria recurva,* is found in this habitat.

One of the few trees in these wet areas is Tasmanian blue gum *Eucalyptus globulus.* As the name implies, these trees are not native; some have been planted, some have germinated naturally. Most *Eucalyptus* species are endemic to Australia.

Fig. 6.7   Cow parsnip

Those which grow along Australia's southern and western edges experience a climate similar to that of central and southern coastal California. Realizing this, local entrepreneurs introduced several eucalyptus species to California in the late nineteenth century, hoping that the trees would grow as fast here as they did in Australia and produce a cash lumber crop in short time. They were not disappointed in the growth rate, but the timber proved too inferior and hard to cut for most purposes. The investors lost their shirts, but California gained a beautiful tree. For the negative aspect of plant and animal introductions, see Chapter 7. (Incidentally, California has reciprocated generously; coastal Monterey pine has been introduced to Australia where it produces better timber in fewer years than in its native habitat.)

Seeps

Wherever they occur, these trickles of fresh water produce strange plant associations. Seeps along ocean-facing bluffs, for example, support species characteristic of the inland grassland, the intertidal zone, and dry roadsides: scarlet pimpernel of the grassland, gumweed, *Grindelia stricta* ssp. *venulosa* of bluff lips, sow thistle of the grassland, annual bulrush, *Scirpus koilolepis* of the area just above the reach of the waves, and the introduced roadside annuals rabbit foot grass, *Polypogon monspeliensis,* and field mustard, *Brassica campestris.* Seeps along the high part of the salt marsh permit freshwater marsh species such as cinquefoil and brass buttons to associate with *Jaumia* of the salt marsh and field mustard of disturbed roadsides and sea rocket of the beaches. These seeps offer valuable experimental opportunities or places to determine which environmental factors are most critical to plant establishment, and under what conditions the limiting factors may be compensated for or otherwise overcome.

7

Man in the Bodega Head Ecosystem

By the chapter title we mean to imply the effect of *modern*, "civilized" man on Bodega Head. Human beings have lived in North America for at least 10,000 years, possibly 40,000. But in the Bodega Head area, at least, native peoples probably exerted very little disturbing influence on the ecosystem, certainly by comparison with ourselves. Most likely they did not live in large communities, they were mobile, their population density was low, they did not set herds of domestic animals to graze the area, they did not introduce exotic plant and animal species, they did not mine or lumber extensively, they did not produce potentially toxic by-products from large-scale industrial or sewage plants. These are the activities of

Fig. 7.1   Excavation made for nuclear reactor ("hole in the head").

modern man, and whatever effect they have had on Bodega
Head is quite recent—within the last 200 years.

The Effect of Plant and Animal Introductions

When exotic organisms are introduced to an area which
they have not occupied before, generally only two results
may accrue. One possibility is that an introduced species may
simply not become established in any of the new habitats
available to it. It is incapable of competing with any of the
native species. Its niche has already been filled, and it is not
flexible enough to occupy any alternative niche, or alternative niches are not there. This is probably the fate of most

introduced organisms. Considering the mobility of people over the past hundred years, surely it is surprising that so little of the world's flora and fauna has become cosmopolitan— unless one assumes that many introductions have failed.

A second result might be that the introduction becomes —if we can apply the word to both plants and animals—a weed, in the sense of an organism that occupies areas heavily disturbed by man. Weed species are normally pioneers (Chapter 2), and as such have great dispersal and reproductive potential as well as very generalized requirements for existence. They are, in short, opportunists. Because disturbed or early successional niches are continuously becoming available, especially in the years since the rise of technological man, weed-type organisms are likely to be more successfully introduced than others. The tendency, given no further disturbance, is for native species that are adapted to the natural environment to become reestablished and for the exotic species to be displaced from the particular area. But because of the continual production, ubiquity of new disturbed sites throughout the world, an exotic species may well sustain itself in its new range. The house mouse (*Mus musculus*) belongs in this category, and so do a number of annual plants, the starling, the English sparrow, and the Norway rat.

THE CASE OF THE DISAPPEARING MOUSE

As we mentioned in Chapter 2, the house mouse is present at Bodega Head, but is restricted to the disturbed areas near buildings; it has never been trapped in the tall, thick, "undisturbed" grassland where shrews, voles, and deer mice abound. Why has this species of European grasslands mouse been excluded from the Bodega grassland? Theoretically there are a number of explanations. It may be that in his size, nesting behavior, and food habits, he occupies a niche similar to that occupied by the deer mouse (say) and thus competes with him. But the deer mouse, which has evolved in ecosystems like that of Bodega Head, may be better adapted and have a competitive edge. For example, it may be better adapted to avoid natural predators such as the marsh hawk, barn owl, weasel, or sparrow hawk. Exactly how the deer mice may be better adapted—be it that they run faster, that their fur offers better camouflage, or that their burrows are a bit deeper—could only be determined by experiments comparing the two animals. Whatever the exact mechanism, the deer mice would under those circumstances consistently tend to reach adulthood, mate and leave more offspring than the house mice.

Another hypothetical explanation for the exclusion may be competition for space. If there are only so many nesting sites available, if a certain number of square feet of space are required by a breeding pair, and if the native mice can maintain themselves there, then there may be no room for an introduced species, which nests in a similar fashion, to gain a toe-hold. This seems to be the reason that a population of house mice recently became extinct on an island about 60 miles south of Bodega Head. William Z. Lidicker, Jr. (1966), a zoologist at the University of California at Berkeley, became interested in a small, fifty-five acre island in San Francisco Bay. Brooks Island was relatively undisturbed and isolated from the mainland. There were no buildings or roads on the island, and none of the native, mainland mammals. The vegetation was very similar to that of Bodega Head grassland, dominated by annual grasses, with occasional shrubs (the shrubs were coyote bush rather than lupine; consequently, we can guess that the soil was sandier and poorer in nutrients than Bodega Head grassland soil).

Sometime prior to 1957, the house mouse had been introduced to Brooks Island. No one knew exactly when or how the introduction came about, but periodic visits by sev-

eral zoologists during the 1950s revealed a house mouse population of high density, scattered throughout the island. Here
was a switch: in the absence of native mammals such as deer
mice and voles, the introduced mouse was not restricted to
disturbed sites. Lidicker set out to study this mouse population and—as is true with many advances in science—he was
accompanied by a great stroke of luck. In October of 1958 he
began checking population size and distribution of the mouse
by an extensive network of live traps, similar to those used by
us. He continued to trap for more than a year, until November of 1959. At that time he had to abandon his project, not
because of the usual reasons graduate students stop work, but
because the house mouse had become extinct on Brooks Island. In October of 1958 his trapping data indicated there
were 300 mice per acre, but by December of 1958 that density
had fallen sharply to 100 per acre, and it continued to decline
gradually until it reached 0 in November of 1959. What had
caused the extinction?

Rainfall and forage production had been normal for the
year. Since the mice feed heavily on seeds, he kept track of
seed numbers in the topsoil for the entire period and found
no systematic decline in seed numbers. He found no symptoms of disease, but did notice unusually low levels of fat content. The levels were not so low as to indicate malnutrition,
however; hence, drought, famine, or disease were not the
cause. Potential predators of the mouse included barn owl,
short-eared owl, red-tailed hawk, marsh hawk, sparrow hawk,
Norway rat, and garter snake (all are also present on Bodega
Head, except the short-eared owl). But their population numbers seemed stable—in fact, only the short-eared owl was a
year-long resident and could have been considered a heavy
predator—so the decline was not associated with unusually
heavy predation.

But there was the case of the meadow vole. In the sum-

mer of 1958, just a few months before Lidicker began his
study, meadow voles were introduced to the island. It was an
accidental introduction; how the animals got there is not
known, but their date of entry is sure. During the same period
of time that house mouse population size crashed from 300
per acre to 0, vole population size boomed from 1 to 180 per
acre. Was the rise of one the cause of the fall of the other?
Lidicker pointed out that the two have quite distinct food
habits and are not ordinarily considered to be competitors;
but both utilize similar nesting sites. What happens in nature
when the two encounter each other over the same territory?
To find out, he placed breeding pairs of vole and mouse in
the same large cage; baffles throughout prevented contacts
from being too frequent. Then he sat back to watch. Of fifty-
six encounters (involving many mice and voles) between the
two, the vole won out 88% of the time and this usually resulted in the house mouse being forcibly evicted from her
nest. There were occasional bites, but no direct fatalities as a
result of the encounters. Nevertheless, Lidicker felt that encounters in nature might result in three things: eviction of the
mouse, upset of lactation, and upset of pregnancy. At best, the
house mouse would be under duress; at worst, she would be
deprived of nests and young. (A similar series of experiments
conducted by Karl DeLong [1966], also of Berkeley, showed
a similar result; he placed a pregnant mouse in an artificial burrow and an adult vole in a cage connecting with the burrow,
and of eight trials, each lasting four weeks, only one litter was
successfully raised.)

Lidicker was careful not to ascribe the entire extinction
to the vole. At low densities, for example, the mice may not
have been successful in finding mates. But he thought his field
and experimental observations revealed the nature of the competition between house mouse and native fauna. Had there
been disturbed sites on the island, the mouse may have been

able to maintain itself in them; in their absence, it became
extinct.

THE CASE OF THE ANNUAL GRASSLAND

From the best historical records available, plant ecologists feel certain that the aboriginal grasslands of California,
whether those of the coast or of the central valley, were
dominated by perennial plants. One of the most important
perennials was certainly needlegrass (*Stipa* species), a bunch
grass. Today, needlegrass is rare in California (it is present on
Brooks Island) and absent on Bodega; in its place are a number of annual grasses, most of them introduced from the Mediterranean.

What caused the change? It's difficult to be sure, but
experiments conducted in grasslands of Washington, Arizona,
and New Meixco seem to show that grazing pressure is the
first step. Heavy grazing by cattle or sheep has two effects: it
opens the ground up by denuding pockets of it, or by tram-
pling; and it puts pressure on the perennials to regain their
ground cover by germination from seed, rather than by
vegetative growth of rhizomes, stolons, or tillars. And under
that pressure, the introduced annual—which must start each
year from seed—is no longer at a disadvantage in competing
with perennials. Prior to grazing, the perennials may completely cover the ground with dense foliage, and they have
extensive root systems which soon drain surface soil dry;
annual seedlings would have difficulty becoming established
under those conditions. But when grazing opens up areas in
which seeds can germinate and develop, areas where both
annual and perennial start from seed, it is possible that the
annuals will grow faster, and the perennial will be at a disadvantage.

The work of Grant Harris (1967), from Washington
State University, clearly shows what can happen if the an-

nual's growth rate is indeed faster than the perennial's. He studied seedling competition between bluebunch wheatgrass, a perennial, and cheatgrass, an annual. About a century ago, bluebunch wheatgrass dominated the intermountain grasslands just as thoroughly as needlegrass dominated those of California. In the middle of the nineteenth century, cheatgrass was accidentally introduced from Europe, and from that time to the present, graziers have noticed an enormous increase in abundance of cheatgrass and an equally impressive decrease in bluebunch wheatgrass.

When Dr. Harris grew seedlings of the two species, both alone and mixed together, he noticed that the presence of cheatgrass greatly reduced growth and survival of bluebunch wheatgrass. He thought this might be due to root competition for moisture. He planted seeds of each in long glass tubes which were filled with soil and inserted in the field, level with surrounding soil. Every month the tubes were lifted and depth of rooting measured. He found that both species germinated at the same time in fall, but that during the winter the roots of cheatgrass grew much faster than those of bluebunch wheatgrass. At the start of rapid shoot growth, in April, cheatgrass roots had penetrated down 90 cm, in contrast to 20 cm for wheatgrass. Cheatgrass was thus able to extract water from a greater part of the soil profile than bluebunch wheatgrass. Not only that, but when the upper soil became dry in early summer, only wheatgrass suffered, for water still remained available at deeper depths where only cheatgrass roots had reached. Cheatgrass each year increases in density by producing more seeds and seedlings, and each year this results in greater competition for moisture in the upper soil and greater stress on wheatgrass.

A similar competition for moisture may be going on in the Bodega Head grassland, and may explain the abundance of annuals and the rarity of perennials. There are only two species of perennial grass in the grassland, *Bromus carinatus* and *Poa scabrella*, and they occupy only 2% of the ground in summer. Annual, introduced grasses, in contrast, are diverse and together cover 30% of the ground. Herbs make up another 30% of ground cover, and many of these are also introduced annuals.

Now that part of the Head, the Marine Preserve, has been fenced, and grazing and human disturbance kept to a minimum, are any changes in the vegetation noticeable? It may be too early to tell. Furthermore, we have no accurate records of what the plant cover was like before the area became a preserve. Mrs. Gaffney, the owner prior to 1962, claims that the summer-spring show of flowers is more spectacular now than it was when the area was grazed. This may indicate a resurgence of native species, for the color show is dominated by native annuals. By comparing aerial photographs taken in 1970 and 1954, we have been able to detect a three-fold increase in the area dominated by lupine shrubs (see Chapter 2). But when exactly did lupine shrubs begin to increase? Was it during the time that grazing (mainly of horses) was still going on, between 1954 and 1962, or was it after the property became a preserve in 1962? Because we don't have the data or aerial photographs for the intervening period, we shall never know. However, from the data collected in detailed vegetation surveys that we are now conducting, the future changes—if they occur—will be detected and studied.

The Effect of Compaction: The Roadside Flora

Soil of roadsides and paths is compacted; many of the native plants apparently find the hard surface unsuitable for germination and seedling establishment, and their roots cannot grow very well in the poorly aerated subsoil. Rain water penetrates far less readily here than it does into loose, granular grassland soil, and hence, less is retained. It does not require many trips—even on foot—for the soil to be compacted and the flora to change. But the floras along roadsides and paths in grassland are not quite the same, possibly because of factors other than compaction, such as degree of openness of the habitat.

In spring, for example, roadsides are dominated by tall shoots of yellow-flowered field mustard, *Brassica campestris*, but footpaths have little or no field mustard and instead commonly support wild radish, *Raphanus sativa*. In summer, roadsides are dominated by several species: wild oats (*Avena* species); black mustard, *Brassica nigra*, with an inflorescence more branched and smaller flowers than field mustard; brass buttons; lotus, *Lotus corniculatus*; yellow sweet clover, *Melitotus indicus*; and rabbit foot grass, *Polypogon monspeliensis*, are all common. In late summer and fall several tall herbs pre-

Fig. 7.2   Filaree

dominate: horseweed, *Conyza canadensis;* sweet fennel, *Foe-niculum vulgare,* with licorice-smelling, finely dissected foliage; and curley dock, *Rumex crispus.* Other roadside plants at that time include white sweet clover, *Melitotus albus,* and California blackberry, *Rubus vitifolius.*

Characteristic footpath species, growing in the path or right at its edge, are the dandelion-like hairy cat's ear, *Hypochoeris radicata,* which is in flower nearly every month of the year; the prickly, prostrate, matted *Cardionema ramosissimum;* plantain, *Plantago lanceolata;* and the small-flowered but colorful red maids, *Calandrinia ciliata.* Further off the path, but nevertheless in a relatively disturbed zone, are wild radish, *Raphanus sativus,* with flowers of white, yellow, or blue; an annual lupine, *Lupinus bicolor;* one species of filaree, *Erodium cicutarium,* with blue flowers and long-beaked fruits; yarrow, *Achillea borealis;* wild heliotrope, *Phacelia distans,* with filmy, fragile blue flowers; johnny-tuck, *Orthocarpus erianthus,* in dense patches, supposedly a root parasite of surrounding herbs; the foul-smelling skunk weed, *Navarretia squarrosa,* common in summer; and the beautiful blue-eyed grass, *Sisyrhinchium bellum.*

The small, brightly-colored house finch, or linnet, *Carpodacus mexicanus,* common in disturbed areas, is a colorful match for gaudy, bright colored weed flowers.

The Effect of Effluents (and of the Pacific Gas and Electric Company)

The effect of man-made effluents on ecosystems has been discussed in so many newspaper articles, speeches in Congress, and books ranging from the most popular to the most technical journals, that we are not going to belabor or repeat the subject here. The theme which runs through every such story—whether it has to do with human sewage, waste materials from industrial plants, pesticide residues, smoke particles and gasses such as sulfur dioxide from burning coal, or thermal effects of discharge water from power plants—is that the effluent's effect gets magnified in passing through the ecosystem. A pesticide applied at a level too low to affect man directly in a given area may eventually affect the photosynthesis and productivity of his crop plants, or the reproduction of game animals such as fish and birds, or the growth of noxious plants and animals resulting in lower land values.

Although some effluents taken in by organisms can be excreted or detoxified, others remain unaltered and accumulate in tissue; in this case, they are passed through the food chain ultimately to reach animals far removed from the source of the effluent, or—in the case of pesticides—far removed from the target organisms. The results of a spray program carried on at Clear Lake, California, so well summarized by Dr. Robert Rudd (1966, 1970), illustrates this point.

Clear Lake was an excellent recreation area, except for annoying summer gnats. The insecticide DDD (a close relative of DDT) was mixed into lake water in 1949, 1954, and 1957 to kill gnat larvae on the lake bottom. The amount of insecticide used was quite small—only enough to achieve a concentration of 0.02 parts per million—but larval kills were 99%. In 1954, a few months after the second application, over 100 dead western grebes (the same species as at Bodega) were found around the lake. There was no external evidence of disease. After the 1957 application, grebes again perished in large numbers; this time fatty tissue was removed from some of them and analyzed for DDD. The tissue was found to contain 1,600 ppm DDD, a concentration some 80,000 times that of the lake water!

In the Clear Lake ecosystem, grebes are "top carnivores"; they form the end of a series of predators and their prey (see

Fig. 7.3). When components of that food chain leading to grebes were analyzed, their content of DDD was as follows: water = 0.02 ppm; plankton = 5 ppm; plankton feeders = 15 ppm; fish = 1,000 ppm; grebes = 1,600 ppm. Each component of the chain had ingested large masses of the component below it (see, again, Fig. 2.36), but had not modified or excreted the exotic new chemical, so a chain of accumulation resulted. The level of DDD concentration may not have been toxic to anything except gnat larvae at the start of the food chain, but it was at the end: in 1949, there were 1,000 pairs of nesting grebes around the lake, with many young; in 1961, there were only 32 pairs and no young.

Similar magnifications of residues can occur on land. A very elegant field demonstration of the effect of DDT on

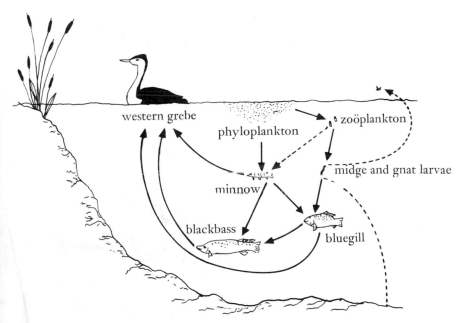

Fig. 7.3 Simplified representation of the food chain in Clear Lake, California. The midge and gnat larvae were the target organisms of added DDD, but all the organisms above accumulated DDD in their tissues. Courtesy R. L. Rudd 1969.

bird mortality was compiled by Wurster and others in New England a few years ago. The town of Hanover, New Hampshire, had sprayed its elms annually with DDT for fifteen years in order to control dutch elm disease. The fungus that causes the disease is carried by a bark beetle from tree to tree as he feeds on live twigs and breeds in dead or dying wood. DDT seems effective in killing the beetle, thus preventing the spread of the fungus. Norwich, Vermont, just one mile west, had never sprayed its trees.

In April of 1963, Hanover applied its annual spray and Wurster kept track of the bird populations in both towns for the next three months. In Hanover he collected 151 dead or dying birds, representing 34 species, during that period; in Norwich he collected 10 dead birds, representing 7 species. One of the dead species found in both towns was the robin; analysis showed that all 5 dead robins in Norwich contained less than 5 ppm DDT in their brain tissue, but 20 of the 27 dead robins in Hanover contained more than 50 ppm. By feeding caged, healthy birds different amounts of DDT in their diet, they duplicated the symptoms of dying birds in Hanover, and from autopsies concluded that 50 ppm DDT in brain tissue was indeed a lethal concentration. Ovary tissue of dead birds contained even more DDT—about ten times the brain concentrations.

How had the birds acquired such large concentrations? Ground feeders, such as robins, picked it up from ground prey such as earthworms who in turn had ingested it in soil, which in turn had accumulated and retained the residues when they washed off tree foliage in rain water. Crown feeders, like the myrtle warbler, ironically picked it up by eating large amounts of living bark beetles—the target of the spray. These beetles had DDT on and in them, but not a lethal dose; when the beetles were devoured in large numbers by crown feeders, the birds accumulated enough chemical residue that they died.

Another abnormality that pesticides can induce is an eggshell too thin to allow the embryo inside to develop and eventually hatch. This effect may partly be responsible for the decline in numbers of a great many fish-eating birds during the past twenty to twenty-five years: ospreys, pelicans, egrets, herons, cormorants, and others. During 1969 and 1970 the ornithologist Frank Gress studied a colony of brown pelicans on Anacapa Island, off the California coast at Santa Barbara. He found that the 1970 breeding season was disastrous for these birds: of all the eggs laid by 550 breeding pairs, only one hatched; the rest had shells so thin and brittle they broke long before the embryos inside had developed into chicks. He analyzed yolk tissues from broken eggs, and found extraordinarily high amounts of insecticide residues. Lawrence Blus, of the Patuxent Wildlife Research Center, has reported similar findings for the brown pelican on the Atlantic and Gulf coasts. There, eggshell thickness and weight have declined an average of 13% since 1957.

The brown pelican is common at Bodega at certain times of the year. Immature birds are often seen there, but quite conclusive tagging evidence has shown that these immature birds are all transients and actually hatched in central Mexico.* Brown pelicans are not known to breed successfully north of Anacapa, and as pesticides work their way south, southern rookeries may decline as well.

Reports by field naturalists in England (Ratcliffe) and Wisconsin (Hickey and Anderson), and many other places, support the contention that eggshell thickness has been declining since 1945 for a number of predatory birds, in parallel with their declining numbers. These species include peregrine falcon, sparrow hawk, golden eagle, bald eagle, osprey, and herring gull. These studies were conducted by measuring eggs in nature and a great many dated ones in museum collections.

* S. G. Herman: personal communication.

The study by Hickey and Anderson (1968), for example, showed that shell thickness of peregrine falcon eggs in California has declined 19% since 1946, while from 1900 to 1946 the thickness had remained very constant. The peregrine falcon was once a common predator at Bodega, and occupied a well-known nesting site on the cliffs overlooking the ocean. The bird has now passed into virtual extinction in the western United States. The period 1940–46 is an interesting one, for it marked the beginning of widespread DDT use; the chemist Paul Muller first developed DDT in 1939 (he later was awarded the Nobel Prize for that work).

These findings, though strong circumstantial evidence, are still only correlations. Controlled feeding of birds is necessary to scientifically determine whether there is a direct relationship between pesticide levels in their food, eggshell thickness, and mortality. Such controlled experiments have only recently been conducted.

For instance, a group of United States Department of Agriculture scientists in Maryland, led by Joel Bitman (Bitman et al. 1969), fed Japanese quail a diet laced with 0 to 100 ppm DDT for forty-five days; another Maryland group, led by Heath (1969), fed mallard ducks a diet enriched with 10 ppm DDT or DDD for two years; and a third Maryland group (Porter and Wiemeyer 1969) treated sparrow hawks with 0, 8, or 16 ppm DDT and Dieldrin for two years. The result of all experiments showed that insecticide levels as low as 8 ppm significantly reduced eggshell thickness, reduced the number of successful hatchings, and increased chick mortality. Such birds in nature could be expected to eat food naturally contaminated with 10 ppm of insecticide, so the levels used in these tests were not unnaturally high. The mallard and sparrow hawk studies indicate that exposure to insecticides for long periods of time produces cumulative damage: the percent of surviving mallard chicks, as compared to a control group, was only 51% when parents had been fed 10 ppm DDD for one year, but it was down to 25% when parents had been fed 10 ppm DDD for two years.

The worst aspect of the pesticide problem is the global prevalence of these substances. Dr. R. W. Risebrough and co-workers (1968) have shown that pesticides, carried by the northeast trade winds, contributed to a decline in the reproduction of the Bermuda petrel—a bird that does not come within hundreds of miles of places where pesticides are actually used. Sladen and others (1966) discovered only recently that pesticides were accumulating in the food chains of Antarctica. There does not seem to be any spot on the globe, no matter how remote, which has been untouched by pesticides.

Even if the use of DDT and related chlorinated hydrocarbons is prohibited by law, as it is in several states, the chemicals will still be circulating in the ecosystem for some time because they are broken down so slowly by living tissue; Rudd expects them to remain abundant enough to cause biological damage for decades, and perhaps indefinitely. Through its effects on pollinating insects, on fisheries, and on other biological resources, it threatens to cost humanity billions of dollars annually over many years. And there are many non-pesticide effluents discharged into bays and harbors, whose effects on organisms have not been studied at all, let alone as carefully as the effects of DDT.

THE PG&E STORY

The southern tip of Bodega Head was owned by Pacific Gas and Electric Company (PG&E), where the company had hoped, at one time, to construct an atomic power plant. This hope conflicted with the plans and desires of private property owners, fishermen, educators, and conservationists; it was supported, on the other hand, by officials of Sonoma County, by the Public Utilities Commission, and by land developers. This type of conflict over land use is becoming more and more common in this century. The one concerning Bodega Head extends over six years, from 1958 to 1964, marked by bitterness, acrimony and often intemperance, and leaving scars which have yet to heal.

Early in 1958, PG&E officially approached the Sonoma County Board of Supervisors with a proposal to construct a power plant on Bodega Head. In fact, however, officials of the company had been sounding out members of the board individually, since 1956. Information was quite vague about the plant at this time: PG&E stated that the plant "might" be atomic powered, and that it would be located either on Horseshoe Cove or further south on the headlands. By a split vote, the Board granted a use permit to PG&E for some 225 acres which the company had already purchased privately.

In 1961, PG&E made its plans more definite and more public. According to the *San Francisco Chronicle* of 29 June 1961: "The Pacific Gas and Electric Company yesterday announced details of its plan to build a $61 million atomic power plant at Bodega Bay. . . . It will be by far the largest such plant in the United States and will generate enough electricity to serve a city of 500,000 persons." Horseshoe Cove was bypassed in favor of the more southern headlands, and construction was scheduled to begin in August of 1962, to be completed in 1965. In addition, an access road in the tidelands along the east side of the Head was to be constructed. PG&E President Norman R. Sutherland stated there would be no radiation danger, and that wastes would be treated and prevented from contaminating the ocean or the air. Sea water would, however, be used to cool the plant's condenser: 250,000 gallons per minute would be taken in from the east side of the Head and, after circulating through the plant, be dumped back into the ocean on the west side, some 18°F warmer.

This announcement by PG&E immediately produced a long series of pro and con statements by local groups and individuals. Conservationists, supported by the marine biologist Dr. Joel Hedgpeth at the University of the Pacific and David Pesonen, spokesman for the Northern California Association to Preserve Bodega Head and Harbor, did not accept the assurance of no radiation hazard, thought that the heated water would have some effect on intertidal organisms, and were not convinced that the site was geologically sound (earthquake-proof). Some who depended on fishing for an income were worried about the possible effect on fish populations. Others, who came to Bodega Bay to retire or to avoid more crowded cities, thought of human populations and objected to the stimulated development of housing and industry which would undoubtedly accompany construction of the plant. Many others, of course, thought of an expanded tax base and other economic assets sure to result from plant construction, and they favored the proposal.

The University of California, Berkeley, Committee on the Marine Biological Laboratory, which had been searching for a laboratory site near San Francisco, was anything but pleased by the announcement, even though it had known something of PG&E plans since 1957. The Committee was especially upset because it had come to the conclusion that Horseshoe Cove was the most ideal site for the future laboratory. In a 1960 report to then U.C. Berkeley Chancellor, Glenn Seaborg, the Committee reiterated its reasons for selecting Bodega Head as its first choice: diversity of habitats, relatively undisturbed compared to other central California areas, and close to the campuses it was meant to serve. The only serious shortcoming ". . . is now the unpredictable degree of ecological instability resulting from the activities of PG&E. At best, the ecological effects might be very minor. . . . Bluntly stated, a unique class A site for a marine facility is being exploited for power production." But this committee report was not to be released to the public for two years.

PG&E had several hurdles to pass before actual construction could begin: the Army Corps of Engineers had to approve construction of the tidelands access road; the California Public Utilities Commission (PUC) had to approve plant construction; and the United States Atomic Energy Commission (AEC) had to approve licensing of the plant.

The first battle, over the tidelands road, took five months, from November, 1961, to March, 1962. The Army Corps of Engineers' proceedings culminated in a marathon eight-hour public hearing at Bodega Bay in February. Although PG&E opponents insisted that the road would only serve the interests of the power plant, and that it might cause filling of the channel and added dredging problems to the county, the Army Corps of Engineers agreed with those who said it would improve access to tidelands for recreation purposes, and it approved road construction. Rose Gaffney, who had opposed the power plant from the beginning, owned 65 acres along the proposed access route, and she refused to sell the land at condemnation evaluation. In April, by a 9–3 jury vote, she was forced to sell at a price of $1,000 per acre, roughly twice the condemnation evaluation.

Coinciding with the road decision, the PUC opened their hearings. PG&E engineers said the reactor would be located a mile from the San Andreas Fault zone, and would be bedded on granite, surrounded by concrete, and designed very conservatively. PG&E also announced plans to develop the recreational value of their land; for example, they planned to reconstruct the old Russian fort at Campbell Cove, and to finance archeological explorations. Opponents had their days on May 21 and 22. According to the *San Francisco Chronicle* of 22 May 1962: "A heavily outnumbered contingent from the Pacific Gas and Electric Company traded shots with an army of aroused nature lovers as the battle of Bodega Head opened here yesterday. At least 150 demonstrative scientists, fishermen, politicians, and other angry citizens appeared. . . .

Witnesses accused the big company of a multitude of sins ranging from collusion with public officials to contamination of marine life." Insinuations that PG&E applied pressure on the University of California to keep its scientists out of the fight, and that President Clark Kerr and the Board of Regents yielded to that pressure, bordered on slander. It does seem odd that some Berkeley biologists did not speak out more strongly or that the University itself did not take at least a mild stand in behalf of its own interests with respect to the marine laboratory. But in testimony on May 22, Dr. A. Starker Leopold, assistant to the U.C. Berkeley Chancellor, and an eminent conservation biologist, put the University firmly on the sidelines, saying flatly that it ". . . neither supports nor opposes . . ." the PG&E plans.

Testimony from biologists was confusing. A fisheries professor from Humboldt said the 18°F rise in outfall water from the power plant would cause no problems with fishing or fish populations. Others suggested that existing data were insufficient to conclude that marine life would be adversely affected by power plant operations. Dr. Hedgpeth, on the other extreme, insisted that such data did exist, and that there would be major ecological consequences. It may be that Dr. Hedgpeth was closer to the truth, for a similar plant built by PG&E in 1963 near Eureka, California, has turned out to be ". . . the dirtiest of the nation's power reactors . . ." by 1971 standards of radiation emission (*Science*, news and notes, 18 June 1971).

An anonymous folk song, celebrating the battle of Bodega Head, was not long in coming. Thirteen stanzas appeared in the *Sebastopol Times* of 2 August 1962; here are two of them:

> The Indians lived on Bodega
> Their middens are there by the sea—
> The Indians are gone, remembered by song,
> Will this happen to you and me?

> What will become of Bodega,
> Dillon Beach and Tomales Bay,
> When PG&E puts their stuff out to sea—
> What will happen to you and me?

On November 10, PUC granted permission to PG&E "to construct, install, operate, and maintain . . ." the power plant. Later that month, U.C. Berkeley biochemistry professor James B. Neilands declared that an important university faculty report on the power plant had been "suppressed" so that PUC could not include it in their deliberations. He asked PUC to reopen hearings. The report he referred to was the 1960 Marine Biological Laboratory Committee report that had lain dormant in the Chancellor's office for two years. The report was given full play in the newspapers, but PUC declined to reopen hearings. And, despite several other calls to reopen hearings during the next two years, PUC never reopened them.

PG&E attempted the third hurdle, the AEC, by applying for licensing in December of 1962. This was actually a two-step hurdle, for the application was examined by two groups in the AEC: the Advisory Committee on Reactor Safeguards, and the Division of Reactor Licensing.

In May, 1963, the Safeguards Committee announced conditional approval of PG&E plans. "Tentative exploration indicates that the reactor and turbine buildings will not be located on an active fault zone. If this point is established, the design criteria for the plant are adequate." But this approval was challenged with very heavy artillery a week later by an independent group called the Northern California Association to Preserve Bodega Head and Harbor. According to the *Sebastopol Times* of 9 May 1963, that group made four charges: 1) The reactor was within a quarter of a mile of the San Andreas Fault zone, not a mile away, as PG&E claimed, and federal regulations prohibit construction of a reactor within a quarter of a mile of an active fault. 2) Association experts "have found evidence of an active earthquake fault right through the proposed reactor site." which is a branch from the San Andreas Fault. 3) The "granite bedrock" claimed by PG&E to underlie the reactor site is actually badly fractured and crushed quartz diorite, which has the stability of clay, not of hard rock. 4) The summary report given the AEC by PG&E repressed or altered initial reports by PG&E staff which showed the unsuitability of the location.

Interior Secretary Stewart Udall wrote a letter to AEC Chairman Seaborg later that month, expressing "grave concern" about the Bodega project's possible effects on coast fisheries from reactor accidents, on the new Point Reyes National Seashore, and on local people in general from seismic hazards. In October, a U.S. Geological Survey team sent to Bodega Head by the AEC concluded that there was indeed a fault on the reactor site. Although they could not determine if the fault were active or not, one of the seismologists said, "Acceptance of the Bodega Head as a safe reactor site will establish a precedent that will make it exceedingly difficult to reject any proposed future site on the grounds of extreme earthquake risk." (*San Francisco News Call Bulletin,* 4 October 1963.)

Still PG&E pushed ahead in its presentation of the Bodega Head site for reactor construction. A PG&E seismologist held a news conference on October 7 in the then sixty-one-foot-deep circular pit being dug to hold the reactor. He pointed out the fracture which ran all across the bottom of the pit and up most of the south side, but thought it "very unlikely" that it was active.

The AEC was silent for a full year. Then, on 26 October 1964, it released two reports. The Advisory Committee on Reactor Safeguards concluded that the plant could be built "without undue hazard to the health and safety of the public." But the Division of Reactor Licensing concluded that "Bodega Head is not a suitable location for the proposed nuclear power plant at the present state of our knowledge." The implications were that the AEC approved of the site in principle, but would not license it. A hearing, whose date was not set, would precede a final decision by the AEC. Governor Brown of California publicly urged PG&E to seek another site.

In a story carried by the *San Francisco Examiner* of 31 October 1964, PG&E President Robert H. Gerdes acknowledged that he had withdrawn the licensing application. He insisted that the location was safe, but that "We would be the last to desire to build a plant with any substantial doubt existing as to public safety." Although a commendable position to take, it seems to have been rather late, considering the serious allegations made against the PG&E plant during the previous four years. Perhaps it was taken so late because of the large amount of development money PG&E had already put into the site: $4 million since 1958, including purchase of land, construction of a road, and the digging of a pit 73 feet deep and 142 feet in diameter.

Possibly, if all the hearings PG&E went through had been held prior to expenditure of development money, and without the cost of land raising, PG&E might have yielded to the mounting evidence sooner. It is difficult enough to place a dollar value on conservation and weigh it against the economic benefits of development; when $4 million dollars of development money have already been spent, impartiality on the part of the developer is expecting too much. Significantly, PG&E has changed its tactics. In 1970 PG&E acquired an option to buy land along the coast about fifteen miles north of Santa Cruz, California, with the ultimate aim of constructing as many as six nuclear power plants. Unlike Bodega Head, this land is heavily disturbed pasture and scrub land, merging

inland to redwood and Douglas fir. Nevertheless, before buy-
ing the land PG&E hired the environmental planning firm of
Eckbo, Dean, Austin, and Williams of San Francisco to com-
pile a $250,000 impact study: to predict, if possible, the effect
of the power plants on the natural environment and on the
economic status of the area. In 1970–71, two of us (Craig and
Drysdale) participated in that study. At this time, PG&E has
still not decided whether to proceed or not with development,
but it is in a better position to decide than it was ten years ago
at Bodega Head.

It may be possible, however, that each developer must
evolve in his responsiveness along the same bumpy road taken
by PG&E, and that for a while each developer will not learn
from the experiences of others. During the period 1967–69,
a bitter fight was waged over a proposed nuclear power plant
on Cayuga Lake in New York. The participants were different, mainly New York State Electric and Gas Corporation
(NYSE&G) and Cornell University scientists, but the mis-
takes were the same as at Bodega Head. As summarized by
Dorothy Nelkin (1971), in her case study of the Cayuga Lake
story, NYSE&G purchased $700,000 worth of land for the
site and developed it for construction at an additional cost
of $1 million before final constructional permits had been
granted by the AEC, and in the midst of controversy, studies,
and counter-studies. In April, 1969 NYSE&G decided to indefinitely postpone the AEC application, pending further
studies which may require two or three years. And, as at
Bodega Head, biological testimony was sought, and found,
which supported both sides of the issue.

We believe that planning decisions could be made more
smoothly and more honestly if non-partisan, biologically-
oriented consulting teams were available—to developers and
conservationists alike—to conduct pertinent research and rec-
ommend guidelines. Investigative teams were formed and used

at Bodega Head and Cayuga Lake, but they were fielded by special interest groups, and were from the start suspect. Persons outside the academic world are insufficiently aware how easily some employees of a university can be influenced or intimidated. Full-time, independent teams are required, and such teams are at present almost nonexistent. Land use consultant companies are invariably engineering or architecture oriented in staff and outlook; such companies cannot provide biological information except by piece-meal subcontracting. Despite a listing of 500 land use consulting companies in the 1971–72 Pollution Control Directory compiled by the journal *Environmental Science and Technology* (September 1971), we are aware of only a few which are staffed by ecologists. There is a need for such teams, and they may well provide a useful home for the increasing number of ecology Ph.D.s.*

The Federal Water Pollution Control Administration recently conducted an environment-oriented survey of California coastal power plants. The survey was led by Robert Zeller, a civil engineer, and Robert Rulifson, a fisheries biologist (1970). They concluded that water intake and outfall locations are placed to meet engineering and Coast Guard regulations, but without thought for the protection of environmental values. Outfalls often raised offshore water temperature 4°F or more above ambient over an area of 100 acres, 15 feet thick. Although biological changes appear to be minor in such heated water, "The lack of definitive pre- and post-construction field data . . . makes it impossible to document the exact changes that have occurred . . . or to evaluate the effects of those changes." We hope this book is a step in the direction of providing "pre-construction field data."

Considering our chosen standard of living, and the pop-

---

* California is trying to field its own teams; recent reports on Elkhorn Slough (Browning, 1972) and Bolinas Lagoon (Giguere, 1970) are good steps in this direction. The creation of the Environmental Protection Agency in 1970 has also brought the national government into this area.

ulation density, the coastal ecosystem cannot remain unutilized and undisturbed. But how much disturbance is acceptable? How flexible is the ecosystem, and how much can it bend without breaking? How can it be rationally manipulated to simultaneously satisfy all needs: industry, fishing, conservation, rubbish pit, and recreation? The information in this book would hardly suffice even to begin answering those questions, but in pointing out the nature of the food chains and communities, it offers the future experimentalists—who will come up with the answers—a collection of starting points and organisms to deal with.

The Native Indians of Bodega Head

Many of the Indian groups of California did not form close-knit tribes but rather were composed of many bands, each band occupying a relatively small territory, and related bands sharing a number of cultural traits, including a common language. Some of these groups in turn are classified together as what anthropologists call the California Culture, in distinction to the Northwest Coast Culture (with its striking totem poles) and to the Pueblo Culture of the southwest (with associated apartment-like living). The California Culture may represent one of the oldest aboriginal groups in the United States.

One of the California Culture groups was the Coast Miwok. This group's territory included all of present-day Marin County plus southern parts of Sonoma and Napa Counties (Fig. 7.4). The Coast Miwok were closely related to other Miwok groups, which lived inland in Lake, San Joaquin, Calaveras, Tuolumne, Stanislaus, Marisposa, Madera, and Merced Counties; however, these inland relatives will not be discussed here. It is well established that the Coast Miwok were divided into two dialect subgroups: the western, or Bodega dialect, and the southern, or Marin dialect.

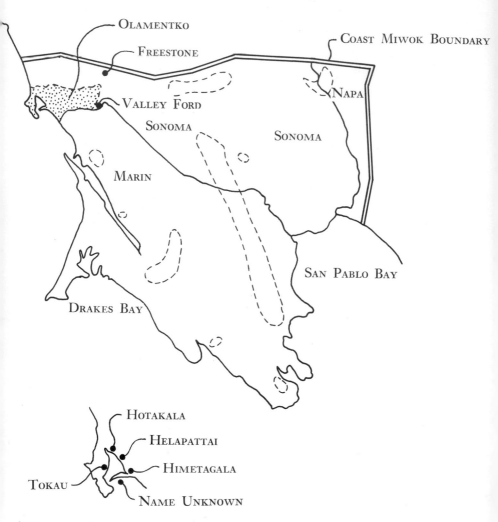

Fig. 7.4 The area occupied by the Coast Miwok, ca. 1770. The Olamentko dialect was restricted to the dotted region about Bodega Harbor. The Hukueko dialect occupied the rest of the area enclosed by the double line, and regions of maximum population density are enclosed by dotted lines. The inset of Bodega Harbor shows the locations of Olamentko villages. (Redrawn from Colley 1970 and Barrett 1908.)

The Bodega dialect, or Olamentko, had a territory which was rather small, about 30–35 square miles (Fig. 7.4). It was bounded on the north by Salmon Creek; on the east by a line due south from near the present town of Freestone to where Valley Ford is today; on the south by the Estero Americano Creek, the present dividing line between Sonoma and Marin Counties; and on the west by the Pacific Ocean.

The Marin dialect, or Hukueko, occupied the rest of the Coast Miwok territory, although there is evidence that they were not scattered uniformly throughout that area (Fig. 7.4). Taking both Olamentko and Hukueko together, the most careful estimate of population size prior to disturbance by whites (ca. 1770) is 1500 to 3000. Given a territory of roughly 1000 square miles, this is a population density of 1.5–3.0 persons per square mile. Although this is low by today's standards (California now supports 120 people per square mile), it was relatively high for those times: according to the research by A. L. Kroeber (in Heizer and Whipple, 1951), about 70% of the state had an Indian population density less than one person per square mile.

The anthropologist S. A. Barrett (1908) carefully collected extensive field data at the turn of the twentieth century, and he concluded that the Olamentko had seven permanent villages. Five of these were on the edges of Bodega Harbor (Fig. 7.4, inset): one, whose name could not be discovered, was situated on the sandy spit now called Doran Beach; Himetagala, at the extreme southeastern corner of the harbor; Helapattai, just south of the present town of Bodega Bay; Hotakala, just north of that town; and Tokau, on the sandy shore near the southwestern corner of the harbor, very close to the present entrance gate to the Marine Laboratory. Bodega Head itself has supplied many Indian artifacts, dating back several centuries. Mrs. Gaffney, the previous owner of the Marine Laboratory property, has an extensive collection of these at her home in Salmon Creek.

The Olamentko people were loosely organized in the villages, with the family being the strongest social unit. One of the villages (possibly Himetagala) served as a chief village in which ceremonies were held and in which the chief resided. Chiefs were generally older men who possessed great experience, tact, speaking ability, and dignity; they held no real power other than that accorded them through persuasion and respect. The Olamentko were not warlike. Disputes over a number of subjects often were settled by payment in goods by one party to another, but occasionally they led to armed conflict. In these engagements, the opposing bands fired arrows at each other from some distance; hence, deaths were few. In fact, the battles were stopped when even one important man on either side was seriously injured or killed.

This Bodega group exploited the biological resources of their habitat with great thoroughness. As one European observer put it, ". . . they eat all creatures that swim in the water, all that fly through the air, and all that creep, crawl, or walk upon the earth, with perhaps a dozen exceptions."

Plant food included many of the species still found on Bodega Head. The bulbs of soap plant were baked like potatoes; they were also used raw as a soap or as fish poison. Bulbs of blue dicks were also eaten. Young, curled tops of bracken fern were roasted, and the roots and rhizomes boiled. Leaves of miner's lettuce, cow parsnip (in the spring), clover, lupine, and goosefoot were either eaten raw or boiled like spinach. Flowers of some species (clover, mallow) were eaten raw. A number of intertidal and subtidal algae (including kelps such as sea palm) were used for their salt content or cooked in ovens and made into cakes.

The animals taken included deer, brush rabbits, jackrabbits, quail, ducks, geese, and snails from the land, and dolphin, porpoise, sea lion, harbor seal, octopus, abalone, chitons, clams, barnacles, mussels, snails, crabs, sea anemones, and sea urchins from the sea. It is estimated that 20% of their daily 100 mg protein intake came from shellfish. Shellfish meat was, in addition, dried and traded to inland groups. The archeologist E. W. Gifford (1949) thought that as much as 50% of the shellfish meat caught at Bodega was carried inland, but most workers think it was 30% or less. Washington clams were dug up from the mudflats and eaten, and the shells were used as money for trade with northern groups such as the Pomo. The Olamentko claimed that this clam was found nowhere else, and as far as Indians further north were concerned, this may have been true. Today, its distribution ranges from Humboldt Bay to San Quintin Bay in Baja California, and we have collected it in the harbor and also a few specimens on the open coast just south of the Russian River. Has the species expanded its range northward in the past 200 years? It seems more likely that the clam populations just north of Bodega Harbor were too sparse for the Indians to find them, rather than that the species has expanded its range.

A good picture of the Indians' animal food can be gained by examination of their refuse heaps, or middens. A large one near the village site of Himetagala was excavated in 1949 by University of California scientists. The midden was about 300 feet long, 150 feet wide, and 13 feet deep. Sections of the mound were removed at successive 1-foot intervals and analyzed for animal remains. About 33% of the material was shell. Table 7.1 shows the proportion of this shell material by species. The California mussel was by far the most abundant species, making up over 50% of all shell at every depth. There were several significant changes in species content with depth (that is, as the deposits got older): the Olympia oyster, bentnosed clam, bay mussel, purple olive snail, turban snail, and barnacle peaked in utilization (abundance) in the middle

Organisms recovered: common (and scientific) names	DEPTH (IN FEET FROM THE SURFACE)				
	0–1	2–3	3–4	4–5	6–7
California mussel (*Mytilus californianus*)	54.00	43.90	53.20	37.60	66.80
horse clam (*Tresus nuttallii*)	16.90	24.60	11.00	27.40	15.50
Washington clam (*Saxidomus nuttalli*)	<u>10.50</u>	1.40	8.50	3.50	.50
bent-nosed clam (*Macoma nasuta*)	3.30	<u>10.40</u>	8.10	7.90	4.30
rock cockle (*Protothaca staminea*)	2.70	9.50	6.80	<u>10.10</u>	3.80
Olympia oyster (*Ostrea lurida*)	.60	.70	.80	<u>1.40</u>	.70
bay mussel (*Mytilus edulis*)	.03	.10	<u>.30</u>	.20	.06
basket cockle (*Clinocardium nuttallii*)	—	1.20	1.10	1.80	<u>3.80</u>
limplet (*Acmaea* spp.)	trace	trace	trace	.06	.05
purple olive snail (*Olivella biplicata*)	.20	.10	<u>.50</u>	.03	.01
turban snail (*Tegula* sp.)	.30	.40	<u>.90</u>	.40	.30
rock snail (*Thais* sp.)	<u>.80</u>	—	—	—	.06
barnacle (*Balanus* sp.)	2.10	3.50	2.80	<u>5.20</u>	1.50
crabs	.50	.50	.50	<u>1.10</u>	.40
black chiton (*Katharina tunicata*)	<u>1.80</u>	.10	.30	.09	.20
purple sea urchin (*Strongylocentrotus purpuratus*)	.30	<u>.40</u>	<u>.40</u>	.20	<u>.40</u>
terrestrial lined snail (*Helminthoglypta arrosa*)	.03	—	—	—	—
others and unidentified	5.94	3.70	4.80	3.02	1.62
total	100.00	100.00	100.00	100.00	100.00

Table 7.1 Composition of a shellmound near the village site of Himetagala on Bodega Harbor. The percent of total shell by species is given for different depths in the mound. Peaks of abundance are underlined. (Adapted from Greengo, 1951.)

period; the Washington clam, rock snail, and black chiton peaked in the most recent period; the basket cockle was used less and less as the deposits become more recent.

Other Coast Miwok shellmounds reveal a similar composition. McGeein and Mueller, for example, carefully analyzed the contents of a mound just north of Sausalito. That mound contained many of the shellfish species found in the Bodega mound: bent-nosed clams, mussels, rock snails, barnacles, horse clams, rock cockles, and the Olympia oyster. One major difference they found with depth was a four-fold increase in abundance of the bent-nosed clam as the deposit became more recent. Mammalian bones of deer, dolphin, porpoise, and sea otter were found. Fish bones were found but not identified. Bones of some twenty-three species of birds were recovered: the most abundant were bones of ravens, double-crested cormorants, loons, scoters, old squaws, marbled godwits, hawks, and Canada geese. Bird bones were most abundant in the middle depths (55% of all bones), and declined in old and young deposits (5% of all bones).

Archeologists who examined these middens assumed that the abundance of species in the mounds reflects their past abundance in nature. This is not strictly true, however, for one must also consider that species are utilized in proportion to a work/food ratio. Barnacles, for example, are abundant but small and difficult to collect, and are not utilized to the same extent as mussels, which are larger and require less work for a meal (compare, in Table 7.1, the amount of mussel and barnacle shell at the 0–1 foot depth: 54% vs 2%).

Why did the mound contents fluctuate with time? A recent paper by F. P. Shepard (1964) summarized some evidence to show that sea level has greatly increased over the past 15,000 years. Fifteen thousand years ago, sea level was about 250 feet below what it is today; 6000 years ago, it was 20 feet below present level. Divers along the southern California coast have recovered thousands of small mortars left by the Indians who lived along the shore at least 5000 years ago. As sea level rose, the coast outline changed, and the distribution of open coast, protected harbors, and mudflats changed in parallel. Some of the animals used for food are restricted to open coast (California mussel, rock snail, turban snail, sea urchin), others to protected mud and sand flats (basket cockle, bent-nosed clam, Washington clam), so the distribution of coast habitats surely would have had an impact on Indian food habits. In addition, as sea level rose, the distance between the shore and the villages narrowed, and this could have had an impact on availability of food organisms and on food preferences.

Several investigators have tried to estimate the age of shellmounds using formulas relating to the amount of shellfish needed per day per person, and what volume of shell that would produce in a midden. But the archeologist M. A. Glassow (1967) has shown that such estimates may be considerably (25%) in error. Based on carbon-dating of charcoal, and other evidence, it seems that the central coast mounds are no older than 3500 years before the present (BP). Since older artifacts of man are present in inland California (to about 5000 years BP), this implies that Indians may have migrated to the central California coast from the interior rather recently.

Despite the paucity of ancient artifacts, it seems likely that man has been in California longer than 5000 years. Most archeologists assume that man reached the western hemisphere from Asia via the Bering Strait and migrated slowly southward, first into North America, then to South America. Man-made hunting tools have been recovered from the tip of South America at the Strait of Magellan which were associated with animal bones and charcoal carbon-dated to 11,000 years BP. Man must, therefore, have reached North America considerably earlier than 11,000 BP. He may, for example, have reached California 100,000 years ago. But at present there is no hard evidence to support that hypothesis.

The first encounter the Coast Miwok had with white men occurred in 1595, when Cermeño anchored his ship San Augustin in Drakes Bay and claimed the surrounding land for Spain. Construction of the mission Dolores was begun in San Francisco in 1776; its completion marked the start of a dramatic Indian decline.

The historian C. C. Colley (1970) has recently documented the effect of missionary activity on the Coast Miwok people. Indians brought to the missions found the food inadequate, the living quarters crowded and unsanitary, and the idea of repetitious, regulated workdays abhorrent. Occasional lay visitors have left written comments on ecclesiastical intolerance and ignorance of Indian culture. Poor living conditions, measles and syphilis were largely responsible for a sharp drop in Indian population during the period from 1780 to 1820. For example, a measles epidemic struck the San Francisco area in 1806, and records indicate that 34% to 88% of Indians aged ten or less died, depending on the exact area. The native population in the San Francisco area was virtually extinct by 1816. Other missions opened in Marin and Sonoma Counties, but the missionaries found it difficult to persuade the Olamentko to move inland to them.

Colley concluded, "At the end of the Spanish mission period, the Coast Miwok had the questionable distinction of being the only tribe north of San Francisco Bay that came under the complete and thorough subjugation of the missions." Coast Miwok population dropped from 1500–3000 in 1770 to less than 15 in 1910. In 1908 Barrett wrote that there were not more than four or five full-blood Olamentko left. Of course, other California Indian groups were also decimated during the same period: Indian population in all of California in 1910 was only 12% of what it had been in 1770.

There seems to be little in common between this extinct Indian culture and the industrialized society that has replaced it. The Indian culture lasted thousands of years. How long will ours last? Theirs seems to have been one based on a stable population size. Ours, for the moment at least, is based on growth. Simple mathematics tells us we cannot grow forever; our only basic question is, when do we stop increasing? To answer, "When the quality of life is significantly reduced," is no answer at all, *unless we realize what its quality has been and what it is now*. We hope this book has helped define that past and present quality of life at Bodega Head. Our readers will help determine its future quality.

8

Epilogue

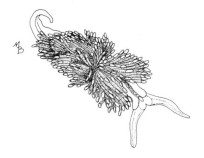

In the present work we have sought to describe a real ecosystem—not an idealized scheme, but rather a detailed case in point. We have presented an inventory of the important and interesting organisms which in fact do live there, and have documented our account with measurements of environmental factors as they actually were recorded on real instruments. A textbook abstraction would perhaps have stressed the means at the expense of the variance—played down the unique and the capricious. When one treats the universe as if averages alone counted, he obtains a very biased view of things. For example, it was lately noted, to the surprise of some ichthyologists at the Bodega Marine Laboratory, that the live-bearing perch were coming into a state of reproductive readiness much earlier in the spring of 1972 than they had the year before; a glance at the temperature data showed that 1971 had an unusually cold spring.

To provide a truly satisfying account—from the point of view of thoroughness—we should have had to provide data on quite a number of additional organisms, especially the smaller and less conspicuous ones. We have said virtually nothing about the bacteria and fungi of the soil. Every animal mentioned has its parasites, and we could have looked at those, too. We have remained silent concerning tiny creatures like mites and protozoans, and have left out the remarkable minute fauna that dwells between sand grains.

We might have examined quite a number of additional features of the Bodega Head ecosystem. How does wind affect the distribution of insects, and, indirectly through pollination, the distribution of plants? We have only mentioned in passing the cycles of nutrients and the problems surrounding trace elements. One could go on and on.

It would have satisfied us very deeply had we been able to place every fact in its proper theoretical perspective. But theory, especially in a vigorous and actively-growing branch of learning, changes through time. Scientists have been considering the relationships between organisms and their environments for centuries. The modern science of ecology, however, was only founded in 1859—in the third chapter of Darwin's masterful *Origin of Species*. And the word "ecology" wasn't invented until ten years later; it remained an obscure, technical term unknown to the general public for another century. Some of the ideas we have applied were introduced into biological thought by Darwin himself: competition and the niche, for example. Others, such as the food chain and pyramids of numbers, were invented during the first half of the present century. Many, such as some thoughts on species diversity, have been taken from actively-developing, current research; and, as can be seen from the many questions for which we can find more than one answer, the explanations are tentative at best. For much of what we have described, the applicable theories have yet to be devised. If the history of science tells us anything, it tells us that ecology will change through time.

So, too, with the ecosystem of Bodega Head: it is anything but static. Fifty years from now, no matter what happens, the flora and fauna and their environments will differ. The area may be given over to housing, or recreation, or industrial development. The preserve may be abandoned, or extended, perhaps used for a different academic purpose, or just left alone. Presence of the Bodega Harbour housing development, begun in earnest in 1973, certainly would lead us to guess that the press of population around the Head will increase several-fold in the near future. Yet even the most undisturbed ecosystem may change spontaneously, and some external influence can scarcely be avoided. Our successors, if any, will have to examine a somewhat different situation, and may not have access to the kind of data we might have provided had we only known. However incomplete our summary of the current Bodega Head ecosystem is, there will never be a more complete summary for this moment in time. Nonetheless, a beginning has been made: it shall be possible to observe what transformations do in fact occur. We reside in a changing world, and must come to grips with change if we are to make ourselves at home in it.

Contents of
Appendixes

Appendix A

Checklist of Vascular Plants
and Vertebrates at Bodega Head

I. **VASCULAR PLANTS.** Notes include scientific and common
names, habitats where found, relative abundance, and time of
peak flowering.

Sphenophyta (Horsetails).
   Equisetum arvense L. Common horsetail. Perennial, native.
   Occasional in seasonally wet, disturbed areas near roads.
   E. hyemale L. var. **robustum**. A. A. Eat. Giant scouring
   rush. Perennial, native. Occasional on wet, shaded banks.
   E. telmateia Ehr. var. **braunii** Milde. Giant horsetail. Pe-
   rennial, native. Common on steep, wet hillsides and in the
   fresh-water marsh.

Pteridophyta (Ferns).
   Aspidiaceae.
      Athyrium felix-femina (L.) Roth. var. **sitchense** Rupr.
      Lady fern. Rare to common on shaded streambanks.
      Polystichum munitum (Kaulf.) Presl. Sword fern.

Perennial, native. Common on steep, wet hillsides.
Polypodiaceae.
> **Polypodium scouleri** Hook. and Gray. Polypody. Perennial, native. Occasional among rocks on grassland hilltops.

Pteridiaceae.
> **Pteridium aquilinum** (L.) Kuhn var. **lanuginosum** (Bong.) Fern. Bracken. Common in the lee of grassland hills.

Salviniaceae.
> **Azolla filiculoides** Lam. Water fern. Native. Occasional in still, fresh water. Only vegetative plants seen.

Coniferophyta (CONIFERS).
Cupressus macrocarpa Hartw. ex Gordon. Monterey cypress. Perennial, planted or escaped. Occasional near the fresh-water marsh, and planted in the dunes and grassland.

Anthophyta (FLOWERING PLANTS).
Dicotyledonaea.
> **Aizoaceae.**
>> **Mesembryanthemum chilense** Mol. Sea fig. Perennial, possibly native. Common at the lip of grassland cliffs and in the splash zone of the intertidal, occasional in the dunes and salt-water marsh, rare on the strand. March–Sept.
>> **M. edule** L. Hottentot fig. Perrenial, introduced. Occasional in the dunes, planted along roadbanks. March–July.

> **Anacardiaceae** (CASHEW FAMILY).
>> **Rhus diversiloba** T. & G. Poison oak. Perennial, native. Common among rocks on grassland hilltops and occasional along shaded gulley.

> **Apocynaceae** (DOGBANE FAMILY).
>> **Vinca major** L. Periwinkle. Perennial, introduced. In a dense strip along one edge of the fresh-water marsh. March–Oct.

> **Berberidaceae** (BARBERRY FAMILY).
>> **Berberis pinnata** Lag. Barberry. Perennial, native. Common on grassland hills away from the coast.

Boraginaceae.
  Amsinckia menziesii (Lehm.) Nels. and Macbr.
  Fiddleneck. Annual, native. Common in grassland.
  April–June.
  A. spectabilis F. & M. Fiddleneck. Annual, native.
  Occasional on established dunes. June–July.
  Cryptantha leiocarpa (F. & M.) Greene. Annual,
  native. Occasional in dunes. June.
  Plagiobothrys tenellus (Nutt.) Gray. Annual,
  native. Occasional to common in disturbed areas.
  April–May.
Caryophyllaceae (PINK FAMILY).
  Cardionema ramosissimum (Weinm.) Nels. &
  Macbr. Perennial, native. Common to abundant in
  grassland footpaths. June–July.
  Cerastium arvense L. Mouse-ear chickweed. An-
  nual, native. Rare in grassland. May.
  Sagina crassicaulis Wats. Pearlwort. Perennial,
  native. Rare in low areas in dunes. April–May.
  Silene gallica L. Windmill pink. Annual intro-
  duced. Occasional in grassland. May–June.
  Spergula arvensis L. Spurrey. Annual, intro-
  duced. Occasional in disturbed areas. March–July.
  Spergularia macrotheca (Hornem.) Heynh.
  Sand-spurrey. Perennial, native. Occasional in
  splash zone of intertidal, also in disturbed areas.
  May–Aug.
  S. rubra (L.) J. & C. Presl. Sand-spurrey. Annual,
  native. Common along roads.
  Stellaria media (L.) Cyr. Chickweed. Annual,
  introduced. Common in grassland, abundant along
  shaded streambank. March–April.
Chenopodiaceae (GOOSEFOOT FAMILY).
  Atriplex patula L. ssp. hastata (L.) Hall and
  Clem. Annual, introduced. Occasional at upper
  edge of salt marsh. Aug.
  A. patula L. ssp. obtusa (Cham.) Hall and Clem.
  Annual, introduced. Rare in salt marsh. Oct.
  Chenopodium album L. Pigweed. Perennial, in-

troduced. Rare in seasonally wet, disturbed areas. July.

C. ambrisoides L. var. vagans (Standl.) Howell. Mexican tea. Annual or perennial, native. Rare in disturbed areas. Aug.–Sept.

C. californicum (Wats.) Wats. Pigweed. Perennial, native. Occasional in grassland and disturbed areas.

Salicornia virginica L. Pickleweed. Perennial, native. Abundant in salt marsh, common in splash zone of intertidal. Oct.

Compositae (SUNFLOWER FAMILY).

Achillea borealis Bong. ssp. arenicola (Hel.) Keck. Yarrow. Perennial, native. Common in disturbed areas. May–Oct.

Agoseris apargioides (Less.) Greene ssp. maritima (Sheld.) Jones. Beach dandelion. Occasional on established dunes. April–Oct.

Ambrosia chamissonis (Less.) Greene. Silver beachweed. Perennial, native. Common on dunes. Leaves with broad to pinnate lobing. July–Aug.

Anaphalis margaritacea (L.) B. & H. Pearly everlasting. Perennial, native. Common on steep, wet hillsides. July–Oct.

Artemisia douglasiana Bess. in Hook. Perennial, native. Rare (but in dense patches) in established dunes, grassland, and disturbed areas. Aug.–Oct.

A. pycnocephala DC. Beach sagewort. Perennial, native. Occasional along lip of grassland bluffs. July–Sept.

Aster chilensis Nees. Perennial, native. Occasional in grassland away from the coast. Sept.

Baccharis pilularis DC. ssp. consanguinea (DC.) Wolf. Coyote bush. Perennial, native. Occasional in established dunes and along roads. Sept.–Oct.

Carduus pycnocephalus L. Italian thistle. Annual, introduced. Rare to occasional in disturbed areas. May–June.

Centaurea solstitialis L. Star thistle. Annual, introduced. Rare along roads. Sept.

Chrysanthemum segetum L. Corn chrysanthemum. Annual, introduced. Rare along roads. June–July.

Cichorium intybus L. Chicory. Perennial, introduced. Rare along roads. Aug.

Cirsium andrewsii (Gray) Jeps. Thistle. Perennial, native. Occasional in grassland. May–July.

C. occidentale (Nutt.) Jeps. Thistle. Perennial, native. Occasional on established dunes. May–July.

C. vulgare (Savi) Ten. Bull thistle. Biennial, introduced. Common in grassland and along shaded, wet banks. July–Oct.

Conyza canadensis (L.) Cronq. Horseweed. Annual, introduced. Common to abundant along roads. Sept.–Oct.

Cotula coronopifolia L. Brass buttons. Perennial, introduced. Common in the fresh-water marsh and in seasonally wet, disturbed areas; rare in the upper part of salt marsh. March–Oct.

Erechtites arguta (A. Rich.) DC. Fireweed. Annual, introduced. Occasional in disturbed areas. July–Aug.

E. prenanthoides (A. Rich.) DC. Fireweed. Annual, introduced. Occasional in the fresh-water marsh, common along steep, wet hillsides, sometimes along roads.

Erigeron glaucus Ker. Seaside daisy. Perennial, native. Occasional in established dunes and near grassland bluffs. June–Sept.

Eriophyllum lanatum (Pursh) Forbes var. **arachnoideum** (Fisch. and AveLall.) Jeps. Golden yarrow. Perennial, native. Rare to occasional in dunes. May.

E. staechadifolium Lag. Golden yarrow. Perennial, native. Common along lip of grassland bluffs. July–Sept.

Evax sparsiflora (Gray) Jeps. Annual, native. Occasional in stabilized dunes. April.

Gnaphalium chilense Spreng. Cudweed. Annual

or biennial, native. Occasional to common in grassland and along roads. May–Oct.

G. chilense Spreng. var. **confertifolium** Greene. Cudweed. Annual, or biennial, native. Rare to occasional in grassland and along roads. June.

G. luteo-album L. Cudweed. Perennial, introduced. Rare in seasonally wet depressions. Nov.

G. purpureum Locc. Cudweed. Annual or biennial, native. Rare along roads. May.

Grindelia stricta DC. ssp. **venulosa** (Jeps.) Keck. Gunweed. Perennial, native. Occasional along roads and down grassland bluffs. May–Sept.

Haplopappus ericoides (Less.) H. & A. Perennial, native. Occasional on established dunes. Aug.–Oct.

Hypochoeris radicata L. Hairy cat's ear. Perennial, introduced. Common in disturbed areas. April–Oct.

Jaumia carnosa (Less.) Gray. Perennial, native. Occasional to common in salt marsh and in splash zone of intertidal. Sept.–Oct.

Lasthenia chrysostoma (F. & M.) Greene. Goldfields. Annual, native. Common to abundant in grassland. May–June.

L. minor DC. Ferris. Goldfields. Annual, native. Common in grassland footpaths, rare in undisturbed grassland. March–April.

Layia platyglossa (F. & M.) Gray. Tidy tips. Occasional to common in grassland and along footpaths through it. Ray petals all yellow. May–July.

Madia sativa Mol. Coast tarweed. Annual, native. Common in grassland, occasional along roads. May–Oct.

Malacothrix californica DC. Annual, native. Rare in dunes. April–May.

Picris echioides L. Ox tongue. Perennial, introduced. Rare to occasional along roads. May–Sept.

Senecio vulgaris L. Common groundsel. Annual, introduced. Rare to occasional in dunes. April–May.

Silybum marianum (L.) Gaertn. Milk thistle.

Annual or biennial, introduced. Occasional in grassland. April–June.

Solidago californica Nutt. California goldenrod. Perennial, native. Rare along shaded streambanks. Sept.

Sonchus asper L. Sow thistle. Annual, introduced. Common to abundant in grassland, occasional along roads and in seeps down grassland bluffs. April–Sept.

S. oleraceus L. Sow thistle. Annual, introduced. Distribution and flowering as with **S. asper.**

Wyethia angustifolia (DC.) Nutt. Perennial, native. Rare along roads. April.

Convolvulaceae (MORNING GLORY FAMILY).

Convolvulus occidentalis Gray var. **saxicola** (Eastw.) Howell. Morning glory. Perennial, native. Occasional in grassland. April–May.

Crassulaceae.

Dudleya farinosa (Lindl.) Britt. and Rose. Live forever. Perennial, native. Occasional among rocks on grassland hilltops, down grassland bluffs, and the sides of gulleys. July–Aug.

Cruciferae (MUSTARD FAMILY).

Arabis blepharophylla H. & A. Rock cress. Perennial, native. Common near summits of grassland hills. March–April.

Barbarea orthoceras Ledeb. Winter cress. Biennial or perennial, native. Rare along grassland paths. March–April.

Brassica campestris L. Field mustard. Annual, introduced. Common to abundant along roads. March–April.

B. nigra (L.) Koch. Black mustard. Annual, introduced. Common along roads. May–Aug.

Cakile maritima Scop. Sea rocket. Annual, introduced. Occasional on strand, rare in upper part of salt marsh. March–Oct.

Cardamine oligosperma Nutt. Bitter cress. Annual or biennial, native. Common in grassland. March.

Nasturtium officinale R. Br. Water cress. Peren-

nial, introduced. Common to abundant along
shaded streambanks, occasional in fresh-water
marsh. June–Aug.

Rorippa curvisiliqua (Hook.) Bessey. Yellow
cress. Annual or biennial, native. Rare in disturbed
grassland. June.

R. nasturtium-aquaticum (L.) Britt. & Rendle.
Wild radish. Annual or biennial, introduced. Common in disturbed areas, especially of the grassland. March–July.

Cucurbitaceae (GOURD FAMILY).

Marah fabaceus (Naud.) Dunn. Manroot. Perennial, native. Common in grassland near Mussel
Point and occasional among rocks on grassland
hilltops. March–June.

Cuscutaceae (sometimes included with CONVOLVULA-
CEAE).

Cuscuta salina Engelm. Dodder. Perennial, native.
Occasional parsite on *Salicornia* in salt marsh, but
only in the upper part of the marsh. June.

C. subinclusa Dur. and Hilg. Dodder. Perennial,
native. Occasional on *Jaumia* in upper part of salt
marsh. June.

Frankeniaceae.

Frankenia grandifolia C. & S. Perennial, native.
Occasional in salt marsh. August.

Geraniaceae (GERANIUM FAMILY).

Erodium cicutarium (L.) L'Her. Red stem filaree. Annual, introduced. Common in disturbed
grassland. March–Oct.

E. moschatum (L.) L'Her. White stem filaree.
Annual, introduced. Occasional in disturbed grassland. March–May.

Geranium dissectum L. Geranium. Annual, introduced. Occasional in grassland. April–May.

Hydrophyllaceae (WATERLEAF FAMILY).

Nemophila menziesii H. & A. Baby blue eyes.
Annual, native. Common in grassland. March–
April.

Phacelia californica Chamisso. Perennial, native.

Occasional along lip of grassland bluffs. Not seen in flower.

P. distans Benth. Wild heliotrope. Annual, native. Common along grassland paths. April–Aug.

Labiatae (MINT FAMILY).

Marrubium vulgare L. Horehound. Perennial, introduced. Rare along roads. Nov.

Mentha pulegium L. Pennyroyal. Perennial, introduced. Occasional on established dunes. Sept.

Monardella villosa Benth. var. **fraciscana** (Elmer) Jeps. Pennyroyal. Perennial, native. Rare in grassland. July.

Stachys rigida Nutt. ex Benth. ssp. **quercetorum** (Heller) Epl. Hedgenettle. Perennial, native. Common to abundant in grassland. April–May.

Leguminosae (PEA FAMILY).

Cytissus monspessulanus L. French broom. Perennial, introduced. Rare along roadsides. Sept.

Lotus corniculatus L. Bird's foot trefoil. Perennial, introduced. Common to abundant along roadsides. June–Aug.

L. heermanii (Dur. and Hilg.) Greene var. **eriophorus** (Greene) Ottley. Perennial, native. Occasional in dunes. May–June.

L. subpinnatus Lag. Annual, native. Rare along roads. Aug.

Lupinus arboreus Sims. Lupine. Perennial, native. Common to abundant in grassland somewhat protected from wind and occasional on stabilized dunes. April–Aug.

L. bicolor Lindl. ssp. **umbellatus** (Greene) Dunn. Lupine. Annual, native. Occasional along roads, especially in grassland. April–July.

L. chamissonis Esch. Lupine. Perennial, native. Rare, overall, common on only three dune ridges. April–May.

L. densiflorus Benth. var. **palustris** (Kell.) C. P. Sm. Lupine. Annual, native. Rare to occasional on roadcuts. April–June.

L. nanus Dougl. in Benth. Lupine. Annual, native.

Common in grassland around Horseshoe cove. April–June.

L. variicolor Steud. Lupine. Perennial, native. Occasional along roadsides. April–June.

Medicago polymorpha L. var. **vulgaris** (Benth.) Shinners. Bur medick. Annual, introduced. Occasional to common along roads. April.

Melitotus albus Desr. White sweet clover. Annual or biennial, introduced. Occasional to common along roads. Aug.–Sept.

M. indicus (L.) All. Yellow sweet clover. Annual or biennial, introduced. Occasional along roads. April–July.

Trifolium barbigerum Torr. Colver. Annual, native. Occasional in disturbed areas. April.

T. fucatum Lindl. Clover. Annual, native. Rare along lip of grassland bluffs. May.

T. gracilentum T. & G. Clover. Annual, native. Occasional in grassland and in grassland paths. April–May.

T. macraei H. & A. Clover. Annual, native. Occasional on rocky grassland hilltops. April–May.

T. repens L. White clover. Perennial, introduced. Occasional to common along roads. June.

T. wormskioldii Lehm. Clover. Perennial, native. Rare to occasional in grassland paths and in freshwater marsh. May–July.

Vicia americana Muhl. ssp. **oregana** (Nutt.) Abrams. Vetch. Perennial, native. Occasional to rare in disturbed grassland. April.

V. californica Greene. Vetch. Perennial, native. Occasional in disturbed grassland. May.

V. gigantea Hook. Vetch. Perennial, native. Common on steep, wet hillsides. March–July.

Malvaceae.

Lavatera arborea L. Tree mallow. Perennial, introduced. Rare along roads. July.

Sidalcea malviflora (DC.) Gray ex Benth. ssp. **laciniata** Hitchck. Checker. Perennial, native. Rare along roadsides. April.

Myoporaceae.
> **Myoporum laetum** Forst. Perennial, introduced. Planted near Marine Laboratory in disturbed grassland. March–July.

Myricaceae (SWEET GALE FAMILY).
> **Myrica californica** C. & S. Wax myrtle. Perennial, native. Occasional on steep, wet hillsides. Feb.

Myrtaceae.
> **Eucalyptus globulus** Labil. Tasmanian blue gum. Perennial, introduced. Planted or possibly escaped in wet, shaded gulleys. April–Aug.

Nyctaginaceae (FOUR O'CLOCK FAMILY).
> **Abronia latifolia** Esch. Sand verbena. Perennial, native. Common to occasional along grassland roads and on established dunes. April–Oct.

Onagraceae (EVENING PRIMROSE FAMILY).
> **Camissonia cheiranthifolia** (Hornem. ex Spreng.) Raimann. Evening primrose. Common in dunes. April–Oct.
>
> **C. strigulosa** (Fisch. & Mey.) Raven. Evening primrose. Rare to occasional in dunes. April–May.
>
> **Epilobium adenocaulon** Hausskn. var. **occidentale** Trel. Willow herb. Perennial, native. Occasional along roads and in fresh-water marsh. Aug.–Sept.
>
> **E. watsonii** Barbey var. **franciscanum** (Barbey) Jeps. Willow herb. Perennial, native. Rare along roads. Aug.–Sept.

Oxalidaceae.
> **Oxalis corniculata** L. Wood sorrel. Perennial, introduced. Occasional in grassland. Sept.
>
> **O. pes-caprae** L. Wood sorrel. Perennial, introduced. Rare in shaded, seasonally wet areas. March.

Papaveraceae (POPPY FAMILY).
> **Eschscholzia californica** Cham. California poppy (state flower). Perennial, native. Common to abundant in grassland. March–Oct.
>
> **Platystemon californicus** Benth. Cream cups. An-

nual, native. Common to abundant in grassland. April–July.

Plantaginaceae.

Plantago hookeriana F. & M. var. **californica** (Greene) Poe. Plantain. Annual, native. Occasional along grassland footpaths. April–May.

P. lanceolata L. Plantain. Perennial, introduced. Occasional to common in disturbed areas. April–July.

P. maritima L. var. **californica** (Fern.) Pilg. Plantain. Perennial, native. Common on bare rock in splash zone on intertidal. May–June.

Plumbaginaceae (LEADWORT FAMILY).

Armeria maritima (Mill.) Willd. var. **californica** (Boiss.) Lawr. Sea pink, thrift. Perennial, native. Abundant along lip of grassland bluffs. April–July.

Polemoniaceae (PHLOX FAMILY).

Gilia capitata Sims. var. **chamissonis** (Greene) Grant. Annual, native. Rare in established dunes and disturbed grassland. May.

Linanthus androsaceus (Benth.) Greene. Annual, native. Rare in dunes. April–May.

Navarretia squarrosa (Eschs.) H. & A. Skunkweed. Annual, native. Rare to occasional in disturbed grassland. June–July.

Polygonaceae (BUCKWHEAT FAMILY).

Chorizanthe cuspidata Wats. var. **villosa** (Eastw.) Munz. Annual, native. Occasional along grassland paths. June.

Eriogonum latifolium Sm. Wild buckwheat. Perennial, native. Occasional along lip of grassland bluffs and down their face, also in grassland at some distance inland from the coast. July–Sept.

Polygonum coccineum Muhl. Perennial, native. Rare in the fresh-water marsh. Only vegetative plants seen.

P. patulum Biebst. Knotweed. Annual, introduced. Rare along roads. May.

P. paronychia Cham. & Schlecht. Perennial, native. Rare on beach. March.

Pterostegia drymarioides F. & M. Annual, native. Rare in grassland. April.

Rumex acetosella L. Sheep sorrel. Perennial, introduced. Common in grassland and along footpaths through it. March–July.

R. crispus L. Curly dock. Annual, introduced. Occasional along roads. May–July.

R. pulcher L. Fiddle dock. Perennial, introduced. Occasional along roads. June–July.

Portulacaceae (PURSLANE FAMILY).

Calandrinia ciliata (R. & P.) DC. var. **menziesii** (Hook.) Macbr. Red maids. Annual, native. Common in disturbed areas. March–April.

Montia perfoliata (Donn) Howell. Miner's lettuce. Annual, native. Common to abundant in grassland. March–April.

Primulaceae.

Anagallis arvensis L. Scarlet pimpernel. Annual, introduced. Occasional to common in grassland, in disturbed areas, and rarely in seeps down grassland bluffs. March–Sept.

Ranunculaceae (BUTTERCUP FAMILY).

Delphinium decorum F. & M. Larkspur. Perennial, native. Rare on rocky grassland hilltops. April–May.

Ranunculus californicus Benth. var. **cuneatus** Greene. Buttercup. Perennial, native. Common in grassland and in paths through it. Petals all yellow or white-tipped. March–April.

Rhamnaceae (BUCKTHORN FAMILY).

Rhamnus californica Eschs. ssp. **tomentella** (Benth.) Wolf. Buckthorn. Perennial, native. Occasional in gulleys. Not seen in flower.

Rosaceae (ROSE FAMILY).

Acaena californica Bitter. Perennial, native. Rare in disturbed grassland. May.

Alchemilla occidentalis Nutt. Annual, native. Occasional in stabilized dunes. April.

Fragaria chiloensis (L.) Duchn. Beach strawberry. Perennial, introduced. Rare in dunes and grassland. March.

Horkelia marinensis (Elmer) Crum in Keck. Perennial, native. Rare about rocks on grassland hilltops.

Potentilla egedii Wormsk. var. **grandis** (Rydb.) Howell. Cinquefoil. Perennial, native. Abundant in fresh-water marsh and in seasonally wet depressions, occasional at upper edge of salt marsh. April–July.

Rosa eglanteria L. Eglantine. Perennial, introduced. Occasional in grassland away from coast. Not seen in flower.

Rubus procerus P. J. Muell. Himalaya berry. Perennial, introduced. Along shaded streambanks and in grassland away from the coast; common to occasional. June.

R. spectabilis Pursh. var. **franciscanus** (Rydb.) Howell. Salmon berry. Perrennial, native. Rare to occasional on steep, wet hillsides. March.

R. ursinus C. & S. California blackberry. Perennial, native. Rare in grassland away from coast. April.

R. vitifolius C. & S. California blackberry. Occasional along roads. Perennial, native. Not seen in flower.

Rubiaceae (MADDER FAMILY).

Galium asperrimum Gray. Bedstraw. Perennial, native. Rare in grassland. April.

G. nuttallii Gray. Bedstraw. Perennial, native. Occasional seasonally wet depression. Mature fruit seen in November.

Salicaceae (WILLOW FAMILY).

Salix laevigata Bebb. Willow. Perennial, native. Common in gulleys. March.

S. lasiolepis Benth. Arroyo willow. Perennial, native. Common in gulleys. March.

S. lasiolepis Benth. var. **bigelovii** (Torr.) Bebb.

Willow. Perennial, native. Rare on established dunes. Not seen in flower.

Scrophulariaceae (FIGWORT FAMILY).

Castilleja wrightii Elmer. Paint brush. Perennial, native. Occasional to common on steep, wet hillsides and in seasonally wet (but disturbed) areas, rare on face of grassland bluffs. April–Oct.

Cordylanthus maritimus Nutt. ex Benth. in DC. Bird's beak. Annual, native. Common in salt marsh. July.

Linaria canadensis (L.) Dum.-Cours. Annual, native. Rare to occasional in established dunes. April.

Mimulus aurantiacus Curt. Bush monkey flower. Perennial, native. Common in grassland and along roads away from coast. April–July.

M. guttatus Fisch. ex DC. ssp. **litoralis** Penn. Monkey flower. Perennial, native. Occasional in fresh-water marsh and in seasonally wet areas along roads. April–Oct.

Orthocarpus erianthus Bench. var. **roseus** Gray. Johnny-tuck. Annual, native. Abundant in rare patches in disturbed grassland. April–May.

O. pusillus Benth. Annual, native. Occasional in grassland. April.

Scrophularia californica C. & S. Figwort. Perennial, native. Rare along roads. May.

Veronica americana Schwein. Speedwell. Perennial, native. Occasional to common in fresh-water marsh. July.

Solanaceae (NIGHTSHADE FAMILY).

Solanum nodiflorum Jacq. Nightshade. Annual or perennial, native. Rare to occasional in grassland, in gulleys, and along roads. June–July.

Umbelliferae (CARROT FAMILY).

Angelica hendersonii Coult. and Rose. Perennial, native. Occasional along the face of grassland bluffs. Aug.–Sept.

Anthriscus scandicina (Weber) Mansf. Bur chervil. Annual, introduced. Occasional to rare near roads. May.

Berula erecta (Huds.) Cov. Perennial, introduced. Common in the wetter parts of the fresh-water marsh. Only vegetative plants seen.

Conium maculatum L. Poison hemlock. Biennial, introduced. Occasional in gulleys and along roads. June–July.

Daucus carota L. Queen Anne's lace. Biennial, introduced. Occasional along roads. July–Sept.

D. pusillus Michx. Rattlesnake weed. Annual, native. Rare in dunes. April.

Foeniculum vulgare Mill. Sweet fennel. Biennial or perennial, introduced. Occasional along roads. July–Sept.

Heracleum lanatum Michx. Cow parsnip. Perennial, native. Common on steep, wet hillsides. April.

Hydrocotyle ranunculoides L. Marsh pennywort. Perennial, native. Abundant in the wettest part of the fresh-water marsh. Only vegetative plants seen.

Oenanthe sarmentosa Presl. Perennial, native. Abundant to common in fresh-water marsh, occasional at upper edge of salt marsh. May–Sept.

Sanicula arctopoides. H. & A. Yellow mats. Perennial, native. Occasional in grassland. March.

Monocotyledoneae.

Amaryllidacea.

Allium dichlamydeum Greene. Wild onion. Perennial, native. Rare about grassland hilltops. June.

A. triquetrum L. Wild onion. Perennial, introduced. Rare in seasonally wet area near road. March.

Brodiaea pulchella (Salisb.) Greene. Blue dicks. Perennial, native. Occasional about grassland hilltops. June–July.

Cyperaceae (SEDGE FAMILY).

Carex barbarae Dewey. Sedge. Perennial, native. Occasional along roads. June–July.

C. obnupta Bailey. Sedge. Perennial, native. Occasional to rare in fresh-water marsh. April.

Cyperus eragrostis Lam. Umbrella sedge. Perennial, native. Common in seasonally wet areas and in the fresh-water marsh. Aug.–Sept.

Scirpus americanus Pers. Bulrush. Perennial, native. Common in salt marsh and in brackish ponds. April–June.

S. cernuus Vahl. var. **californicus** (Torr.) Beetle. Bulrush. Annual, native. Rare in dunes. July.

S. koilolepis (Steud.) Gleason. Bulrush. Annual, native. Common in salt marsh, in splash zone of intertidal, and along seeps of bluffs facing the ocean. Aug.

S. microcarpus Presl. Bulrush. Perennial, native. Common in the fresh-water marsh and seasonally wet areas. July.

Gramineae (GRASS FAMILY).

Aira caryophylla L. Hairgrass. Annual, introduced. Common to abundant in grassland. April–June.

Agrostis alba L. Red top. Perennial, introduced. Rare to occasional in swards in the grassland. July.

A. exarata Trin. Bent grass. Rare along shaded streambanks. July.

Ammophila arenaria (L.) Link. Beach grass. Perennial, introduced. Planted in dunes; abundant. July.

Avena barbata Brot. Wild oats. Annual, introduced. Common along roads. May.

Briza maxima L. Quaking grass. Annual, introduced. Rare along roads. June.

B. minor L. Quaking grass. Annual, introduced. Rare along roads. June.

Bromus carinatus H. & M. California brome. Perennial, native. Occasional to common in grassland. May–June.

B. mollis L. Soft chess. Annual, introduced. Rare to occasional in disturbed grassland. June.

B. diandrus Roth. Ripgut. Annual, introduced. Abundant in grassland, occasional in disturbed areas. April.

Calamagrostis nutkaensis (Presl.) Steud. Reed grass. Perennial, native. Common on steep, wet hillsides. Nov.

Cordateria selloana (Schult.) Arch. and Graebn. Pampas grass. Perennial, introduced. Probably planted in grassland near Marine Laboratory; rare. Sept.

Cynosurus echinatus L. Dogtail. Annual, introduced. Rare in grassland. April–May.

Dactylis glomerata L. Orchard grass. Perennial, introduced. Rare along roads. June.

Distichlis spicata (L.) Greene var. **stolonifera** Bettle. Perennial, native. Salt grass. Common to abundant in salt marsh and in splash zone of intertidal. Not seen in flower.

Elymus glaucus Buckl. Rye grass. Perennial, native. Rare in gulleys. July.

E. vancouverensis Vasey. Rye grass. Perennial, native. Occasional in stabilized dunes, along roads, and near the lip of grassland bluffs. Aug.

Festuca dertonensis (All.) Arch. and Graebn. Fescui. Annual, introduced. Common in grassland. May.

F. megalura Nutt. Foxtail fescue. Annual, native. Rare along sandy roadsides. May.

Holcus lanatus L. Velvet grass. Perennial, introduced. Abundant in seasonally wet grassland area, common in fresh-water marsh, rarely in brackish areas. June–Aug.

Hordeum brachyantherum Nevskii. Perennial, native. Occasional in grassland. April–May.

H. depressum (Scribn. and Sm.) Rydb. Wild barley. Annual, native. Common in grassland, occasional along roads. May.

H. leporinum Link. Farmer's foxtail. Annual, introduced. Occasional in disturbed areas. March–June.

Lolium multiflorum Lam. Italian ryegrass. Annual, introduced. Abundant in grassland. May–June.

Poa douglasii Nees. Sand bluegrass. Perennial, native. Occasional on established dunes. March–April.

P. scabrella (Thurb.) Benth. Blue grass. Perennial, native. Occasional in grassland. April.

P. unilateralis Scrib. Blue grass. Perennial, native. Common in grassland, especially around Horseshoe Cove. May.

Polypogon monspeliensis (L.) Desf. Rabbit foot grass. Annual, introduced. Occasional in disturbed areas and in seeps down grassland bluffs, rare along shaded streambanks. June–Sept.

Iridaceae (IRIS FAMILY).

Iris douglasiana Herb. Wild iris. Perennial, native. Common about grassland hilltops, also in grassland away from the coast. March–April.

Sisyrinchium bellum Wats. Blue-eyed grass. Perennial, native. Occasional in disturbed grassland.

Juncaceae (RUSH FAMILY).

Juncus balticus Willd. Rush. Perennial, native. Common in fresh-water marsh and in seasonally wet areas of grassland. March.

J. bolanderi Engelm. Rush. Perennial, native. Rare in seasonally wet areas along roads. Aug.

J. bufonis L. Toad rush. Annual, native. Rare in seasonally wet areas along roads. Aug.

J. effusus L. var. brunneus Engelm. Rush. Perennial, native. Occasional in seasonally wet areas along roads. May–Oct.

J. leseurii Bol. Rush. Perennial, native. Common to abundant in the fresh-water marsh, about fresh-water ponds, and in seasonally wet areas (sometimes along roads). May–Oct.

Luzula subsessilis (Wats.) Buch. Wood rush. Perennial, native. Occasional to common near grassland hilltops. March.

Juncaginaceae (sometimes included with SCHEUCHZERIACEAE).

Triglochin maritima L. Arrow grass. Perennial,

native. Common to occasional in the salt marsh. May–June.

Lemnaceae (DUCKWEED FAMILY).

Lemna valdiviana Phil. Perennial, native. Common in fresh-water ponds. Not seen in flower.

Liliaceae (LILY FAMILY).

Chlorogalum pomeridianum (DC.) Kunth. Soap plant. Perennial, native. Common about grassland hilltops. Not seen in flower.

Fritillaria recurva Benth. Fritillary. Rare in grassland, only one plant ever seen. March.

Typhaceae (CATTAIL FAMILY).

Typha donimgenses Pers. Cattail. Perennial, native. Rare to occasional in seasonally wet areas.

Zosteraceae (EELGRASS FAMILY).

Phyllospadix torreyi Wats. Surf grass. Perennial, native. Common to abundant in the rocky intertidal. Only vegetative plants seen.

Zostera marina L. Eelgrass. Perennial, native. Common in Bodega Harbor, but rarely exposed. Only vegetative plants seen.

II. **VERTEBRATES.** Notes include common and scientific names, habitats where found, relative abundance, and seasonality. All species are native unless specified otherwise. Notes in the bird list, especially, are based almost entirely on our own observations. Identifications were as conclusive as we could make them. Several small song birds were seen but positive identifications were never made, so they are omitted from the list. Perhaps you'd like to make additions to this list yourself (we hope, however, there won't be any subtractions). The list of whales and porpoises includes only those that came on shore during the course of our investigations.

AMPHIBIANS.

California newt. **Taricha torosa** Rathke. Rare in fresh-water marsh. Oct.–June.

Western toad. **Bufo boreas** Bard and Girard. Occasional in grassland. Variable seasonality.

Pacific tree frog. **Hyla regilla** Bard and Girard. Abundant in grassland in spring.

Bull frog. **Rana catesbeiana** Shaw. Introduced. Common in fresh-water ponds and marshes. Variable seasonality.

REPTILES.

Northern alligator lizard. **Gerrhonotus coerleus coerleus** Wiegmann. Common in grassland, occasional in dunes. March–Nov.

Western fence lizard. **Sceloporus occidentalis** Baird and Girard. Occasional in disturbed areas. Feb.–Nov.

Gopher snake. **Pitouphis melanoleucus** Daudin. Rare to occasional in grassland. March–Nov.

Western garter snake. **Thamnophis elegans** Baird and Girard. Abundant in grassland, occasional in dunes. March–Oct.

Common garter snake. **T. sirtalis** L. Occasional in damp areas. March–Oct.

BIRDS.

LOONS.

Common loon. **Gavia immer** Brünnich. Common in harbor. Winter and spring.

Arctic loon. **G. adamsii** Gray. Occasional in harbor. Winter.

GREBES.

Horned grebe. **Podiceps auritus** Zarudny and Loudon. Occasional in harbor. Winter.

Eared grebe. **P. caspicus.** Occasional in harbor. Winter.

Pied-billed grebe. **Podilymbus podiceps** L. Common in harbor.

Western grebe. **Aechmophorus occidentalis** Lawrence. Common in harbor. Winter and spring.

FULMARS.

Fulmar. **Fulmaris glacialis** L. Oceanic. Rare in Bodega area.

Sooty shearwater. **Puffinus griseus** Gmelin. Harbor. Spring and fall.

PELICANS.

White pelican. **Pelecanus erythrorhynchos** Gmelin. Occasional in harbor. Winter.

Brown pelican. **P. occidentalis** L. Common in harbor and rocky shore. Fall and winter.

CORMORANTS.

Pelagic cormorant. **Phalacrocorax pelagicus** Dallas. Occasional on ocean-facing rocks. All year.

Brandt's cormorant. **P. penicillatus** Brandt. Common on ocean-facing rocks. All year.

Double-crested cormorant. **P. auritus** Brandt. In harbor. All year.

SWANS, GEESE, AND DUCKS.

Canada goose. **Branta canadensis** L. Occasional in harbor. Winter.

Black brant. **B. nigricans** Lawrence. Occasional in harbor. Winter.

American widgeon. **Mareca americana** Gmelin. Occasional in harbor. Winter.

Red head. **Aythya americana** Eyton. Occasional in harbor. All year.

Lesser scaup. **A. affinis** Eyton. Common in harbor. Winter.

Golden eye. **Bucephala clangula** Bonaparte. Common in harbor. Winter.

Old squaw. **Clangula hyemalis** L. Rare in harbor. Winter.

Common scoter. **Oidemia nigra** L. Occasional in harbor. Winter.

White-winged scoter. **Melanita deglandi** Bonaparte. Abundant in harbor. All year.

Surf scoter. **M. perspicillata** L. Occasional in harbor. Fall and winter.

Ruddy duck. **Oxyura jamaicensis.** Wilson. Abundant in harbor, winter and spring.

Red-breasted merganser. **Mergus serrator.** L. Occasional in harbor, winter.

HERONS.

Great blue heron. **Ardea herodias** L. Common in harbor, fresh-water marsh, and salt marsh. All year.

Black-crowned night heron. **Nycticorax nycticorax** L. Rare in harbor. Summer.

Common egret. **Casmerodius albus** L. Rare to occasional in harbor. All year.

Snowy egret. **Leucophoyx thula** Thayer & Bangs. Occasional in harbor. All year.

American bittern. **Botaurus lentiginosus** Montagu. Occasional in salt marsh. Winter.

VULTURES.

Turkey vulture. **Cathartes aura** L. Common over grassland and dunes. All year.

HAWKS.

Red-tailed hawk. **Buteo jamaicensis** Gmelin. Occasional over grassland. All year.

Marsh hawk. **Circus cyaneus** L. Common to abundant in grassland, dunes, and fresh-water marsh. All year.

OSPREYS.

Osprey. **Pandion haliaetus** Gmelin. Occasional over the harbor. Most of the year.

FALCONS.

Peregrine falcon. **Falco peregrinus** Tunstail. Rare to extinct in Bodega area.

Sparrow hawk. **F. sparverius** L. Common in grassland. All year.

QUAILS AND PHEASANTS.

California quail. **Lophortyx californicus** Ridgway. Common in grassland.

Ring-necked pheasant. **Phasianus colchicus** L. Occasional in grassland. All year.

RAILS AND COOTS.

Common gallinule. **Gallinula chloropus** Bangs. In fresh-water marsh. All year.

American coot. **Fulica americana** Gmelin. Abundant in harbor and marsh. All year.

OYSTERCATCHERS.

Black oystercatcher. **Haematopus bachmani** Audubon. In rocky intertidal. All year.

PLOVERS, TURNSTONES, AND SURFBIRDS.

Killdeer. **Charadrius vociferus** L. Common in marshes and around pounds. All year.

Semipalmated plover. **C. semipalmatus** Bonaparte. Harbor mudflats.

Snowy plover. **C. alexandrinus** Cassin. Harbor mudflats.

Surfbird. **Aphriza virgata** Gmelin. Occasional on ocean-facing rocks. Fall and spring.

Ruddy turnstone. **Arenaria interpres** L. Occasional on ocean-facing or on harbor rocks. Winter and spring.

Black turnstone. **A. melanocephala** Vigros. Common on ocean-facing rocks or on harbor rocks.

Black-bellied plover. **Squatarola squatarola** L. Harbor mudflats.

Snipes, Sandpipers.

Common snipe. **Capella gallinago** L. Occasional in fresh-water marsh. Winter.

Long-billed curlew. **Numenius americanus** Woodhouse. On rocky shores, tidal flats, and in salt marsh. Fall and winter.

Whimbrel. **N. phaeopus** Latham. Rare in salt marsh and tidal flats.

Wandering tattler. **Heteroscelus incanus** Gmelin. Occasional on rocky shores. Winter.

Willet. **Catoptrophorus semipalmatus** Brewster. Common in harbor and salt marsh. All year.

Greater yellow legs. **Totanus melanoleucus** Gmelin. Occasional in marshes. Winter.

Rock sandpiper. **Erolia ptilocnemis** Ridgway. Occasional on rocky shores. All year.

Least sandpiper. **E. minutilla** Vieillot. Common in harbor. Summer and fall.

Spotted sandpiper. **Actitis macularia** L. Harbor mudflats.

Western sandpiper. **Ereunetes mauri** Cabanis. In harbor. Winter.

Knot. **Calidris canutus** Wilson. Harbor mudflats.

Short-billed dowitcher. **Limnodromus griseus** Say. In harbor.

Dunlin. **Erolia alpina** Vieillot. Common in harbor. Winter.

Marbled godwit. **Limosa fedoa** L. Abundant in harbor and salt marsh. All year.

Sanderling. **Crocethica alba** Dallas. Common in harbor. Winter and spring.

AVOCETS.

American avocet. **Recurvirostra americana** Gmelin. Occasional in harbor. Winter.

PHALAROPES.

Red phalarope. **Phalaropus fulicarius** L. Occasional in harbor. Spring.

Northern phalarope. **Lobipes lobatus** L. Harbor mudflats. Summer.

GULLS AND TERNS.

Western gull. **Larus occidentalis** Audubon. Abundant on all shores. All year.

Glaucous-winged gull. **L. glaucescens** Nauman. Occasional on all shores.

Herring gull. **L. argentatus** Brooks. Common in harbor. All year.

California gull. **L. californicus** Lawrence. Occasional on all shores. Most abundant in winter.

Ring-billed gull. **L. delawarensis** Ord. Common on all shores.

Mew gull. **L. canus** Richardson. Occasional in harbor. Winter.

Bonaparte's gull. **L. philidelphia** Ord. Occasional in harbor. Winter.

Caspian tern. **Hydroprognia caspia** Dallas. Occasional in harbor.

Forster's tern. **Sterna forsteri** Nuttall. Occasional in harbor. Fall and winter.

Arctic tern. **S. paradisaea** Pontoppidan. Occasional in harbor. Winter.

AUKS.

Common murre. **Uria aalge** Pontoppidan. Occasional on rocky shores. Winter.

Pigeon guillemot. **Cepphus columba** Dallas. Common on ocean-facing rocks.

DOVES.

Mourning dove. **Zenaidura macroura** Woodhouse. Occasional in grassland.

OWLS.

Barn owl. **Tyto alba** Bonaparte. Occasional in grassland and gulleys. All year.

Burrowing owl. **Speotyto cunicularia** Molina. Occasional along roads.

Great horned owl. **Bubo virginianus** Ridgway. Rare in grassland. All year.

HUMMINGBIRDS.

Anna's hummingbird. **Calypte anna** Lesson. Disturbed grassland. All year.

KINGFISHERS.

Belted kingfishers. **Megaceryle alcyon** L. Common in harbor. All year.

WOODPECKERS.

Red-shafted flicker. **Colaptes cafer** Gmelin. Occasional in grassland. Winter.

TYRANT FLYCATCHERS.

Western kingbird. **Tyrannus verticalis** Say. Occasional–rare in grassland.

Black phoebe. **Sayornis nigricans** Swainson. Occasional in wet areas.

Say's phoebe. **S. saya** Bonaparte. Grassland.

SWALLOWS.

Barn swallow. **Hirundo rustica** L. Common in grassland and salt marsh. Summer.

Cliff swallow. **Petrochelidon pyrrhonta** Vieillot. Occasional in salt marsh.

Rough-winged swallow. **Stelgidopteryx ruficollis** Vieillot. Grassland. Summer.

JAYS AND CROWS.

Scrub jay. **Aphelocoma coerulescens** Bosc. Occasional in disturbed areas.

Common raven. **Corvus corax** L. Occasional along ocean-facing cliffs.

Crow. **C. brachyrhynchos** Brehm. In disturbed areas.

WRENS.

House wren. **Troglodytes aedon** Vieillot. Occasional in fresh-water marsh.

Long-billed marsh wren. **Telmatodytes palustris** Wilson. Common in fresh-water marsh. All year.

BUSHTITS.

Common bushtit. **Psaltriparus minimus** Townsend. In Monterey Cyprus along road.

Bluebirds.

Western bluebird. **Sialia mexicana** Swainson. Rare in pastures.

Robin. **Turdus migratorius** L. Grassland. Winter.

Shrikes.

Loggerhead shrike. **Lanius ludovicianus** L. Occasional in grassland. Fall and winter.

Starlings.

Starling. **Sturnus vulgarus** L. Introduced. Common in disturbed areas. All year.

Meadowlarks and Blackbirds.

Western meadowlark. **Sturnella neglecta** Audubon. Common in grassland. Winter.

Red-winged blackbird. **Agelaius phoeniceus** L. Occasional in grassland and fresh-water marsh.

Brewer's blackbird. **Euphagus cyanocephalus** Wagler. Abundant in fresh-water marsh, in winter, common in spring and summer.

Brown-headed cowbird. **Molothrus ater** Boddaert. Occasional in disturbed pastures.

Finches and Sparrows.

House finch. **Carpodacus mexicanus** Müller. Occasional around buildings. Summer.

American goldfinch. **Spinus tristis** L. Occasional in grassland. Summer.

Savannah sparrow. **Passerculus sandwichensis** Gmelin. Occasional in grassland.

White-crowned sparrow. **Zonotrichia leucophrys** Forster. Abundant in grassland. All year.

Song sparrow. **Melospiza melodia** Wilson. Common in grassland. All year.

Mammals.

Marsupials.

Opossum. **Didelphis marsupialis** L. Introduced. Shrubby areas around buildings, rare.

Insectivores.

Vagrant shrew. **Sorex vagrans** Baird. Common in grassland, dunes, and marshes.

Broad-handed mole. **Scapanus latimuanus** Grinnell. Occasional in grassland and disturbed areas.

Shrew mole. **Neurotrichus gibbsii** Baird. Rare in grassland and fresh-water marsh.

BATS.

Mouse-eared bat. **Myotis** sp. Kaup. Occasional in gulleys, grassland, and marshes.

Mexican free-tailed bat. **Tadarida brasiliensis** Saussure. Occasional in a variety of habitats.

PRIMATES.

Man. **Homo sapiens** L. Native and introduced. Abundant in disturbed areas, occasional elsewhere.

RABBITS.

Black-tailed jack rabbit. **Lepus californicus** Gray. Common in dunes and grassland bordering dunes.

Brush rabbit. **Sylvilagus bachmani** Waterhouse. Common in lupine-dominated grassland.

RODENTS.

Sonoma chipmunk. **Eutamias sonomae** Grinnell. Rare in established dunes.

Pocket gopher. **Thomomys bottae** Eydoux-Gervais. Occasional in grassland.

Harvest mouse. **Reithrodontomys megalotis** Baird. Common in fresh-water marsh, rare in grassland.

Deer mouse. **Peromyscus maniculatus** Wagner. Abundant in grassland and established dunes.

Vole, or California meadow mouse. **Microtus californicus** Peale. Common to abundant in grassland, dunes, and marshes.

Muskrat. **Ondatra zibethica** L. Occasional in fresh-water marsh.

Norway rat. **Rattus norwegicus** Berkenhout. Introduced. Occasional about wharfs and in disturbed areas.

House mouse. **Mus musculus** L. Introduced. Occasional in disturbed areas.

WHALES AND PORPOISES.

Finbacked whale. **Balaenoptera** sp. Oceanic, rarely washed ashore in Bodega area.

CARNIVORES.

Long-tailed weasel. **Mustela frenata** Lichtenstein. Common in dunes and grassland.

Striped skunk. **Mephitis mephitis** Schreber. Common in grassland.

Raccoon. **Procyon lotor** L. Common in grassland, established dunes, and marshes.

Gray fox. **Urocyon cinereoargenteus** Schreber. Occasional in established dunes, grassland, and fresh-water marsh.

Badger. **Taxidea taxus** Schreber. Rare in grassland.

Coyote. **Canis latrans** Say. Rare in grassland, only signs seen.

Mountain lion. **Felis concolor** L. Rare, one sighting only, near Mussel Point.

House cat, feral cat. **Felis domesticus.** Introduced. Occasional in grassland.

Bob cat. **Lynx rufus** Schreber. Rare in Bodega area.

Sea otter. **Enhydra lutris** L. Rarely seen offshore.

Steller's sea lion. **Eumetopias jubata** Schreber. Occasional on offshore rocks, and in the water.

California sea lion. **Zalophus californianus** Lesson. Common on offshore rocks, and in the water.

Harbor seal. **Phoca vitulina** L. Occasional on offshore rocks, and in the water.

Elephant seal. **Mirounga angustirostris** Gill. Rare, offshore.

RUMINANTS.

Black-tailed deer, or mule deer. **Odocoileus hemionus** Rafinesque. Common in grassland, dunes, and fresh-water marsh.

Appendix B

Additional Details About Methods Used

I. ANEMOMETERS MEASUREMENT OF WIND

Basically, anemometers consist of cups or blades attached to a central axis; as they catch the wind, the whole axis revolves. Expensive versions, costing several hundred dollars, require line voltage (110 volts) and record with a pen on a continuously moving chart the direct reading in wind speed. Some additionally note wind direction with a second pen.

Less expensive models simply count the total revolutions on a small meter, much as the odometer of a car records miles of travel. The experimenter must then convert the reading into miles of wind passed by a simple formula (which differs for each instrument). Average wind speed can be calculated by dividing total miles of wind by the number of hours since the meter was last checked. Some voltage is still required, because the revolutions are tallied by the meter in terms of current being passed:

every time the central shaft revolves, two contacts are brought together and this allows the passage of current. One such model, costing about $150, requires a 45-volt battery. The battery must be protected from salt spray, and the anemometer kept well oiled to protect against corrosion. Smaller anemometers, which can fit in the palm of one's hand, give instantaneous readings only and do not provide totals. They cost as little as $35.

II. GERMINATION VS SALINITY: PROCEDURE

We preferred to make solutions of various salinity by diluting unsterilized, unfiltered sea water collected in Bodega Harbor, so that field conditions were imitated. The effects of artificial, single-salt solutions is often not the same as the effect of the mixed salt which are part of sea water. For some experiments, however, we did use various concentrations of sodium chloride (NaCl) in distilled water.

Seeds are first soaked in beakers of these solutions for a few hours, much as they would lie in a saturated soil of the same salinity during winter rains; then they are placed into petri dishes, the bottoms of which are lined with a thin layer of sterile sand that is saturated with the same solution. If covered and kept at room temperature (about 75° F), evaporation is so low that salinity remains nearly constant for as long as seven days. For experimental periods longer than this, it is necessary to replace the water lost by evaporation or else salinity will increase. The correct amount of water can be added by bringing the dish weight back to what it was at the start of incubation.

III. LIGHT INTENSITY MEASUREMENT

Light meters give direct, instantaneous measurements of light intensity which reaches their "eye" or sensitive element. The element contains a metal (usually cadmium) which emits electrons when exposed to light. The electron flow is translated to movement by a needle on a scale. Depending on the meter, the scale may read in units called foot-candles, lux, lumens, lamberts, or film exposure. One foot candle = 10.76 lux.

One foot-candle is the illumination produced when the light from one candle falls at right angles on a one square foot surface one foot away. The metric equivalent is the lux (the one square meter surface is one meter from the candle). The definition of foot-candle is rather archiac and difficult to visualize, but we have used the unit in our book because it has been widely used in the ecological literature. At noon on a bright day at sea level, light intensity may be 8,000–10,000 ft-c. This corresponds roughly to 1.3 cal/cm²/min. Light meters reading in foot-candles and with a high degree of accuracy cost over $250.

Attention should be paid to the range of wave-lengths to which the light meter is sensitive. The pyranometer we used (see also this Appendix) is sensitive to the range 360–2,500 millimicrons, which includes not only visible light (450–700 millimicrons) but some ultraviolet and considerable infrared. Many light meters are sensitive to the range 300–750 millimicrons, but the one we have used has a filter that restricts the sensitivity just to visible light. And it is visible light which controls the all-important photochemical reaction, plant photosynthesis.

Several inexpensive chemical sensors permit total light intensity to be counted over a period of time. One of these involves the principle that anthracene in solution will polymerize into insoluble crystals of dianthracene when exposed to light; the more light, the more polymerization. Glass bottles filled with the solution can be placed at any position in the community, and some time later collected and the amount of anthracine still in solution determined. Dore (1958) has complied a standard reference table which converts anthracine concentration into foot-candle hours.

IV. Live Trapping Methods, and Estimation of Population Size

We trapped small mammals such as deer mice, voles, and shrews in Sherman traps (see Chapter 2). The traps were arranged along transects in the grassland, or at regular intervals in a grid in the salt marsh. Usually more than one trap was placed at each position. The traps were checked every two hours because some of the animals die in a matter of hours if not released. Each animal caught for the first time was marked by removing two toes, and a code was used so that two different toes were removed at each trap location; in this way, the movement of recaptured animals could be determined.

The mark and recapture technique can be used to estimate population size with the help of standard formulas (Calhoun and Casby 1958; Jolly 1965; Southwood 1966). The simplest formula, called the Lincoln or Petersen Index, requires data from only two trapping periods:

$$\frac{\text{Population size}}{\text{Number trapped first period}} = \frac{\text{Total number trapped second period}}{\text{Number recaptured second period}}$$

More sophisticated variations of this basic formula increase the accuracy of the estimates. Jolly's model, for example, can utilize trapping data for more than two periods. It also takes into account birth and mortality during the trapping period, and compensates for "trap happy" animals who seem to enjoy being caught over and over again.

Several assumptions must be made to use the Lincoln Index: 1) All animals in the population, regardless of age or sex, have an equal chance of being captured; 2) animals which have been trapped once have the same chance of being recaptured as those which have never been trapped (that is, no learning takes place); and 3) the presence of traps has no effect on animal activity. Although we used the Index, the assumptions may not have been correct at Bodega Head; Krebs (1966) found that assumption (2) could not be made for populations of voles near Berkeley.

Larger animals such as skunks and raccoons were trapped in wire mesh Havahart traps. These were baited, in contrast to the Sherman traps. Trials with baited and unbaited Sherman traps give the same number of captures; consequently, we did not bait them. Curiosity, apparently, is a strong character trait of the mice.

V. PEARSON'S COEFFICIENT OF COMMUNITY

Most data presented in this book have not been subjected to elaborate statistical tests. Differences in deposition of salt spray with distance, or in germination with temperature, were large and convincing, and we let them speak for themselves. But zonation of plant species over short distances in the grassland was summarized most simply and dramatically with the help of statistical treatment of sampling data. The computation of Pearson's coefficient of community, and the construction of dendrogram that showed community relationships, were two treatment techniques utilized.

To illustrate the methods and formulas used, assume that we wish to compare the similarity of three communities: A, B, and C. The species composition of each is noted, either from a complete census or from quadrat sampling. In this hypothetical case, there are only a total of four species, identified with the numbers 1–4 in Appendix Table 1. If present in a community, the community is given a value of 2; if absent, a value of 0. In the table, for example, species 1 was absent in community A, present in B and C.

Several sums are then computed: X, the sum of the values of each community; X^2, the sum of the squared values; $(X)^2$, the sum of the values, squared. In the table, for example, X_A is the sum of the values in community A, and is equal to 4. A fourth sum depends upon which two communities are being compared for similarity; only two can be compared at one time. The sum is arrived at by multiplying the value of each species in both communities, then summing the products. In the table, for example, $AB = (0 \times 2) + (2 \times 0) + (2 \times 2) + (0 \times 2) = 4$.

All the sums are used in one coefficient of community formula; if communities A and B were being compared, the formula would be:

$$CC_{A\text{-}B} = \frac{AB - (X_A \times X_B)/n}{\sqrt{(X_A^2 - (X_A)^2/n) \times (X_B^2 - [X_B]^2/n)}}$$

$$= \frac{4 - (4 \times 6)/4}{\sqrt{(8 - {}^{16}\!/_4) \times (12 - {}^{36}\!/_4)}}$$

$$= -0.58$$

where n = total of unique species in all communities, not just the two being compared, or n = 4 in this case. This is an im-

Species and sum	Community A	B	C	Comparison A-B	A-C	B-C
1	0	2	2	0	0	4
2	2	0	0	0	0	0
3	2	2	0	4	0	0
4	0	2	2	0	0	4
X	4	6	4			
X^2	8	12	8			
$(X)^2$	16	36	16			
AB, AC, or BC				4	0	8

Appendix Table 1—Raw data for Pearson's coefficient of community.

portant point, for it illustrates the difference between Jaccard's formula and Pearson's. Jaccard's formula in effect took the two communities being compared out of context; that is, other communities in the series of comparisons are disregarded. But Pearson's formula, in assigning presence/absence values for each species in the whole collection of communities (not just the two being compared), includes more information and magnifies the relationships.

When all other comparisons have been made,

$$CC_{A-C} = -1.00$$
$$CC_{B-C} = +0.58$$

which allows one to conclude that B and C are most alike, A and C most unlike. Communities identical in composition would have a CC of $+1.00$ (or $+100$ if the decimal is removed); those with no species in common would have a CC of -1.00 (or -100).

Graphically, these relations could be summarized as the dendrogram in Appendix Fig. 1 Communities B and C are joined at the $+58$ level, and A is joined to them at its average relationship to B alone or to C alone (average of -100 and $-58 = -79$). Notice that some information is lost in the graph: that A is actually more closely related to B than to C. (References: Kershaw 1964; Snedecor and Cochran 1967; Sokal and Rohlf 1969.)

VI. pH: Alkalinity and Acidity

Soil pH can be measured in the field very inexpensively with the use of a small kit, marketed by Hellige, Inc. (877 Stewart Avenue, Garden City, New York), called the Truog Soil Reaction Kit (less than $10). A bit of soil is placed in a depression plate, an indicator solution is added with a dropper, and the color that the mixture turns is compared with a color chart that equates pH with color (color gradations are given for every 0.5 pH units).

More sensitive (but not necessarily more accurate) measuring devices are operated by battery or line voltage, and have a pair of electrodes that must be lowered into the wet soil sample. Water is usually added in sufficient amount to create a soil paste, in which the soil is glistening wet but free water does not collect at the surface. Soil pH data reported in this book, however, was determined from a 1:1 soil : water extract. Soil was dried at 180°F in an oven for one day; a 50.0 g sample was weighed out,

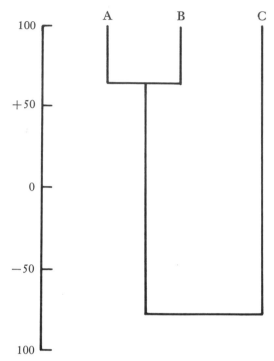

Appendix Figure 1—Dendrogram of communities A, B, and C

mixed with 50 ml of distilled water, and allowed to stand for thirty minutes; then the electrodes were inserted into the supernatant. We found the effect of the added water (over that employed in a paste) was to move the pH about 0.1 unit toward neutrality. The dry soil used is first crushed, mixed, and passed through a 2-mm sieve.

VII. Pyranometer: Measurement of Solar Radiation

The basic sensing element in a pyranometer consists of two flat pieces of metal, one white and the other black, both exposed to the sky beneath a hemisphere of glass. Depending on the amount of sunlight striking the two strips of metal, their temperatures will differ by a great deal or by very little, since the white strip will reflect light and stay cool but the dark one will absorb light and become heated.

This difference in temperature is translated into movement

of an ink pen on chart paper. The paper fits on the outside of a revolving drum. A small, battery-driven motor turns the drum for the desired recording time—a day, a week, or a month. The chart is graduated in units of calories per square centimeter per minute (cal/cm²/min).

On a clear day the ink pen describes a smooth parabolic curve that starts at o at daybreak, reaches a peak of about 1 at noon, and falls back to o at sunset. If fog is present in the morning or late afternoon, the rise in the curve is delayed in the morning and hastened in the afternoon, but the curve is still smooth; if clouds pass during the day, there are sharp drops in the curve when the clouds directly block the sun. The total amount of radiant energy which strikes the ground during the day can be determined by measuring the area under the curve.

We have used a Weather Measure pyranometer, model R-401, which was placed on the roof of the Marine Laboratory, away from all obstructions. This model is sensitive to light of wave-lengths of 360–2,500 millimicrons, which includes all of the visible light plus some ultraviolet (less than 400 millimicrons) and considerable infrared (over 700 millimicrons). It would have been better to have used an instrument sensitive only to visible light, for that is the light utilized by plants in photosynthesis. However, the energy contribution of long wave infrared radiation to the total is small (see Gates 1965). At a price of about $200, this instrument is expensive, but it is essential if one wishes to determine the efficiency of plant productivity (see also this Appendix).

VIII. Plant Productivity Calculations

Plant productivity can be estimated by a reclipping procedure: at the start of a year, all shoot material within a well marked area is clipped back to ground level; at the end of the year, all the shoots which have grown up in the interval are clipped back again, and the clippings dried and weighed. The weight is then converted to caloric content. This may be done directly by taking a sample of the tissue and placing it in a special combustion chamber called a bomb calorimeter. The material is ignited, and the amount of heat released and sensed by the calorimeter is its caloric value. Technically, one calorie is the energy required to raise the temperature of one gram of water one degree centigrade (1.8°F).

Calorimeters are expensive, however, and it is more convenient to estimate the caloric content by referring to caloric values of similar tissue as reported in the literature. Unfortunately, such estimates cannot be very exact for plants. Apart from differences between species, even the same species may exhibit different calorie-per-gram ratios, depending on the growing conditions. Golley (1961) reported that the kilocalorie-per-gram ratio of grass tissue in a Georgia field changed from month to month, from a low 4.2 in early summer to a high of 4.5 in December, and that roots had a different ratio than the shoots when averaged over the year (4.2 for roots, 4.4 for shoots). Vegetation types differed even more: plant tissue in a tropical rain forest averged 3.9 kcal/gram, that in a salt marsh averaged 4.1, and in a pine forest it averaged 4.8.

We did not do any bomb calorimetry, and we selected a ratio used by W. A. Williams (1966) for a pasture in central California: 4.3 kcal/gram. We also estimated root weight as he did, 17% of shoot weight, and added it.

Efficiency of energy conversion by plants for the year period can then be determined as:

$$\text{Efficiency} = \frac{\text{Plant productivity (kcal/square meter)}}{\text{Incoming radiant energy (kcal/square meter)}}$$

where incoming radiant energy is measured by a pyranometer (see also this Appendix).

IX. QUADRAT SAMPLING METHODS

One of the goals of plant ecologists is to categorize all the communities and vegetation types of the world within a uniform classification system. It is important that this goal be realized, for it will result in a cogent summary of the world's plant resources. If the Bodega Head grassland is ever put within such a classification scheme, then ecologists anywhere in the world—who may never have been to Bodega—will be able to compare the grassland to a standard frame of reference and to communities they have seen.

If communities are to be classified, they must be described in some objective fashion. Some characteristics that can be objectively described include: list of species present, species abundance (numbers of plants per unit area), species cover, and the biological features of each species (annual, perennial, tree, shrub, herb, evergreen, deciduous, etc.). It would be tedious to collect this information for an entire community that may cover many acres; instead, only a fraction of the area—often one tenth or less—is sampled with quadrats.

Quadrats are frames of any shape: circles, squares, rectangles, or narrow strips. A line transect, divided into segments, is just a series of rectangular quadrats reduced to the width of a transect tape. The transects are located within a community at random, or at regular intervals, or in a combined arrangement. If they are placed regularly, as they were for this book, then they can be conveniently located at regular, predetermined intervals along transect lines. The transect lines we used cut across the greatest diversity; for example, many went at right angles to the shore rather than parallel to it.

How big should the quadrats be? Big enough to include a representative sample of the plants, small enough to be seen easily by one person standing at its edge. For the grassland, with its large diversity of plant members, we eventually selected a hoop of 1 m²; for the salt marsh, with few members, we used a hoop of ¼ m². There is a more objective method to select optimum quadrat size. One picks a center point, then uses increasingly larger quadrats about that point. As the size increases, more species are included, but there is a point of diminishing returns. Beyond a certain size, the number of added species is quite small. Some select a point of 10% as the dividing line: the optimum quadrat size is at the point where a 10% further increase in size does not increase the number of species encountered by 10%.

How many quadrats are necessary? That depends on how much detail one wants to show. Over the long grassland transects, when we wished to document changes in plant cover over long distances and within large provinces, such as hilltops, hillsides, and protected bottom land, we placed quadrats every 20 m. But to show changes near the bluff edge, we sampled every 5 m. If one were to sample many coastal marshes, however, it would be better to experiment with a large number of quadrats in each zone and determine the point of diminishing returns, just as could be done for quadrat size. Usually, however, 10% or less of the total community area is ultimately included by all the quadrats—however many that may turn out to be. (For more information, see Oosting 1956 and Phillips 1959.)

X. RAIN GAUGES

Rain gauges come in many degrees of complexity and cost. The most expensive ones are tripping bucket rain gauges, which record rainfall as a function of time. As a small bucket fills with rain, it trips a lever, which causes a pen to mark a revolving chart, then empties and begins to fill again. Such an instrument costs several hundred dollars.

The standard rain gauge used by the U.S. Weather Bureau does not record rainfall with time, and must be manually checked and emptied at set intervals of time. It is a hollow metal cylinder whose 8-inch-diameter mouth forms a funnel leading to a smaller-diameter internal cylinder graduated in inches. The gradations are adjusted so that it is possible to read rainfall to within 0.01 inch. If rain fills the inner cylinder, the larger one around it traps the overflow. Total capacity is 20 inches. Standard procedures are that the gauge be mounted on a metal stand and located well away from buildings, trees, and other obstructions. This gauge costs about $75.

Less expensive models, with a smaller capacity and made of plastic, can be purchased for less than $20. We have used plastic models of capacity 11 inches which have an accuracy of 0.1 inch. These can be converted to fog gauges by attaching a 6-inch-high cylinder of screen wire to the top of the lip; condensing fog runs down the wire and into the cylinder just as rain water.

We checked rainfall at weekly intervals, which means that some information was lost: we did not know whether all the rain collected had fallen in one violent storm or as a slow drizzle over the course of many days. These two patterns affect the plants in different ways. If all the rain came at once, the soil probably could not absorb it all; much of it may have run off and been lost to plant roots. If the rain came down slowly, then the soil could have absorbed it all and it would remain available to plant roots. By checking the pyranometer records, however (see this Appendix), we could at least tell how many days of the week had been very cloudy and on which days rain could have fallen.

XI. SALINITY METER

For all our salinity determinations we have used a battery-run meter (Mho-meter, Lab-Line Instruments, Melrose Park,

Illinois), which is expensive ($200) but very convenient to use. The meter is attached to a small cup which is filled with the solution to be tested. Electrodes in the wall measure the conductivity of the solutions; the higher the salt concentration, the greater the conductivity. Conductance (in units called micromhos) is converted to salt concentration by multiplying with an appropriate conversion factor. The value of the conversion factor does depend on the type of salts in solution, but Richards (1954) has shown that the magnitude of change is not great. We have used the following conversion:

parts per million salt = 0.64 × micromhos of conductivity
percent salt = ppm/10,000

Salinity of water samples (for example, ground water in the salt marsh) can be determined directly in the field with this meter, but soil samples must first be extracted and filtered (see soil salinity, this Appendix).

Less expensive methods of determining chloride content (an estimate of total salt) include titration with silver nitrate (see Cox 1967). The concentration of individual ions, such as sodium, magnesium, or chloride can also be determined very accurately with elaborate spectrophotometers (see Cox 1967 and Jackson 1958).

XII. Salt Spray Trapping Methods

One of the techniques we have used to measure salt spray is similar to that used by the plant ecologists Oosting and Billings (1942) on the East Coast. A wood frame, about 7 inches square is mounted on legs just above the soil surface. A square of cheesecloth is tacked to the frame and exposed to the air for a desired period of time; particles of salt spray which strike the cloth remain attached to it. A representative central portion of the cloth is then cut out (without directly touching it) and washed in a beaker of water of known volume. It is assumed that all the salt on the cloth has been washed into solution. We then used a salinity meter (see this Appendix) to measure the salinity of the solution, and work backwards to calculate the rate of salt spray deposition per unit time.

For example, suppose the cut section of cloth measured 10 in^2, and it was washed in 20 ml of distilled water. The conductivity meter registered 1,560 micromhos for the solution.

$$\text{parts per million salt} = 0.64 \times 1,560 = 1,000 \text{ ppm}$$
$$\% \text{ salt in solution} = \text{ppm}/10,000 = 0.1\%$$

Since each ml of water weighs 1 g,

$$\text{weight of salt in solution} = 0.001 \times 20 \text{ g} = 0.02 \text{ g}$$

And since the area of the sample cloth was 10 in^2,

$$\text{weight of salt per square inch} = .02/10 = .002 \text{ g/in}^2.$$

If the cloth was exposed for a day, then the rate of salt deposition was .002 g/in^2/day.

We have also found that a circle of filter paper, taped to the bottom of a petri dish, is an excellent salt trap. Using forceps, the paper is attached with double-stick tape and the petri dish top is taken off for the exposure period. All or a portion of the paper is then removed and washed as with the cheesecloth.

Boyce (1954), another East Coast plant ecologist, experimented with cloth or paper traps cut in the shape of leaves and tied to plants. He thought these traps would replicate natural conditions more closely, and he did find differences in amount of salt trapped by these and by vertical cheesecloth traps. We have tried to use the leaves of sea rocket (*Cakile maritima*) as traps, washing the plant with distilled water at the start of the experiment, then collecting leaves from all over the plant at the end of the exposure period. The leaves are washed again, as with the cheesecloth traps, and salinity measured. Leaf surface area is computed by placing the leaves on photographic paper in the light and measuring the area of their dark images with a small planimeter. But succulent sea rocket leaves have considerable width, which adds more area, and the width is difficult to measure.

XIII. Soil Moisture Determination

We used the most inexpensive method to measure soil moisture: weighing the samples fresh, and after drying (as discussed in the text). There are a number of more expensive methods, however, which do not disturb the soil and allow the same point to be measured again and again.

One popular method involves the use of small gypsum blocks which are buried in soil at desired locations and depths. Two wire leads are embedded parallel to each other in the gypsum; outside the block, they are insulated and run to the surface. When moisture is to be measured, the leads are attached to a battery-run modified ohmmeter. The resistance to the passage of current between the leads in the block depends on the amount of water which fills the porous gypsum; the more water, the less resistance. The amount of water in the blocks reflects the amount of water in the soil, for the two are in equilibrium. The meter scale reads directly in percentage moisture. Unfortunately, salinity can influence the reading. For example, in a soil of 2% salinity, the meter will read 100% moisture regardless of the actual amount of soil water. Ewart and Baver (1950) concluded that gypsum blocks should not be used, because of inaccuracies, when soil salinity is above 0.2%. Some Bodega soils contain this amount, so we did not use the blocks.

A new instrument, called the Aquaprobe (General Scientific Equipment Company, Hamden, Connecticut, about $50), compensates for soil salt and so is more useful. It consists of a handle and meter, with batteries, attached to a 3-foot-long metal tube with pointed tip. Near the tip, inside the tube, is a sensing element. The dial reads in percent available moisture, but we have used it only to compare relative levels of moisture from place to place.

XIV. Soil Salinity: Extraction of Salts

Soil samples, of about 500 g, were carefully removed without unnecessary handling or contamination with nearby soil. They were dried in an oven at about 180°F for twenty-four hours, then a 50.0 g subsample weighed out and mixed with 50 ml of distilled water. The 1 : 1 mixture is stirred briskly, then allowed to equilibrate for thirty minutes. For very sandy soils, it was only necessary to decant the supernatant into the salinity meter cup (see this Appendix), but for soils with more clay, the solutions were filtered.

As discussed in the text, it is important to realize that a 1 : 1 extract is equivalent to a soil with 100% water (dry weight basis). In nature, soils are much drier; consequently, if all the salts in solution at 100% moisture are still in solution at lower percent moisture, the salinity is higher. Depending on the wetness of the soil, roots may sense salt concentrations two or more times that reported in the text. Richards (1954) recommends using a soil paste instead of a 1 : 1 extract.

XV. Soil Texture Determination

The most straightforward way to separate a soil into its sand, silt, and clay components is by sieving it. However, this method is very laborious and it is almost impossible to separate the silt from the clay fraction.

We have used, instead, the hydrometer method discussed in

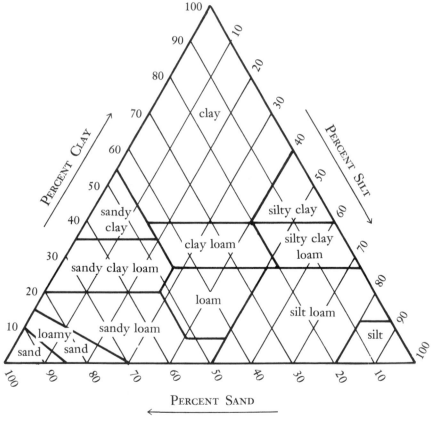

Appendix Figure 2—Soil texture triangle (from Soil Survey Manual 1951)

more detail by Cox (1967) and and Millar et al. (1958). Very generally, a soil sample of about 500 g is collected, dried at about 180°F for twenty-four hours, sieved through a 2-mm sieve to remove gravel, and a 100.0 g subsample weighed out. One liter of tap water is added, and the mixture agitated on a shaking machine for thirty minutes. A pinch of water softener (Calgon brand) helps disperse the soil particles in the water. The solution and sediment is emptied into a 2-foot-tall cylinder and the time noted. After forty-five seconds, a hydrometer (similar to those used to test battery fluid) is inserted in the cylinder and the level to which it sinks is read on a neck scale. From theoretical formulas, it is estimated that at this time all sand has settled; the scale reading is converted to percent sand. After two hours, all silt has settled and a second reading is taken. The percent clay is determined by difference:

$$\% \text{ clay} = 100 - (\% \text{ sand} + \% \text{ silt})$$

The soil texture class can then be assigned by locating the soil on a texture triangle (Appendix Fig. 2). For example, a soil with 60% sand, 30% silt, and 10% clay would be in the sandy loam category.

An additional useful reference is the Soil Survey Manual (Soil Survey Staff 1951).

XVI. SURF GRASS CULTURE METHODS

Clumps of surf grass, on the average 10 cm in diameter, were chipped off the intertidal rocks. Each clump exhibited about fifty to one hundred leaves. Some substrate always remained attached to the rhizomes as the clumps were pried up, and it was allowed to remain attached during laboratory culture. About thirty clumps were gathered at one time, immediately placed in buckets of sea water, and within an hour placed in aquaria under treatment conditions. Seven clumps, of about equal total mass and vigor, were placed in each aquarium.

The aquaria were 50 cm long, 30 cm wide, and 30 cm deep, and constructed of wood and glass, open at the top. They were placed beneath a bank of fluorescent lights timed to provide a twelve-hour day. Light intensity at the water surface was 200 ft-c, and was about half that at plant level beneath 20 cm of water. The tank bottoms were nearly covered by seven clumps, but not crowded, so there was little or no shading of one clump by another.

Sea water was brought to the tanks direct, and largely unchanged, from the ocean via a large pump system. It was continuously added near the bottom of the tanks at a rate of 4.0 liters per minute, and allowed to run out near the top. Air was bubbled through the tanks at a rate of 0.8 liters per minute. The flow of water and air provided some mechanical agitation to the plants, but it was very minor compared to wave action. As a result, debris accumulated on the leaves and the tank bottom, and we found it necessary to drain and wash the tanks and plants once a week to preserve plant vigor.

Leaves of each clump were clipped to a uniform stubble height of 5 cm, the clippings pooled for each aquarium, dried, and weighed. Following a treatment period of about two weeks, the clumps were again clipped to 5 cm and the clippings dried and weighed. Regeneration was calculated as:

$$\%R = \frac{\text{weight of second clipping}}{\text{weight of first clipping}} \times 100$$

Temperature was varied by inserting two aquarium heaters in each tank (12 inch, 250 watt, Aquarium Stock Company, New York, about \$5 each). These were capable of holding the temperature within a range of 3°F of the set temperature. Light was varied by putting different thicknesses of screen over the tank tops. Salinity was varied by regulating the ratio of tap water to ocean water added to each tank.

XVII. TEMPERATURE GRADIENT BAR AND GERMINATION VS TEMPERATURE

A temperature gradient bar similar to the one we used is described by Barbour and Racine (1967). For germination work, an inch-thick bar of aluminum was used. It protruded out the sides of the growth chamber into a hot water bath on one side and a cold water bath on the other. Each bath contains about 200 liters of water; the cold bath is cooled by a ⅙ hp pump attached to cooling coils on the bath bottom, the hot bath by two 250-watt aquarium heaters. With antifreeze in the water, the cold bath can go down to −7°F; a thermoregulator holds the temperature within 1°F. The hot bath can go to 160°F; the heaters have their own built-in regulators which hold the temperature within 3°F. The bar exhibits a gradient of surface temperature between these extremes (or between any other settings).

We have found it convenient to place the seeds in petri dishes whose bottoms are lined with a layer of sterile sand saturated with water. In such a closed dish, the sand acts as a reservoir and the sand needs to be rewetted only periodically (once every two weeks below 70°F, once every two or three days above 95°F). If light inhibits germination, either the dishes may be wrapped in foil or the plexiglass chamber top can be covered.

For seedling growth, $\frac{1}{10}$-inch-thick aluminum can be bent into V-shaped troughs and filled with soil. Partitions every 10 cm prevent roots from growing to warmer or cooler regions, and plant growth will reflect soil temperature differences. The chamber is insulated with 2-inch-thick styrofoam to maintain a uniform air temperature, and a bank of incandescent and fluorescent lights above the plexiglass chamber top provides light.

XVIII. Thermometers

Some thermometers, which register maximum and minimum temperatures sensed since the last time they were set, are moderately priced. The Weather Bureau uses two rugged thermometers, a minimum with alcohol and a maximum with mercury. The alcohol tube encloses a little rod which is drawn back by the contracting meniscus, but is not moved when the temperature rises; its position, then, indicates the lowest temperature sensed since the thermometer was last checked. The maximum thermometer has a constriction which allows mercury to expand through it, but not to pull back; consequently, the mercury column will indicate the maximum temperature sensed.

We have used a Taylor U tube max-min thermometer, which at $10 is only one sixth the cost of the Weather Bureau model. This model utilizes only mercury columns, which push markers to maximum and minimum positions. We have also used dial types which have a two-inch-long sensitive probe at the end of a six-foot-long lead. The dial at the other end of the lead has three hands, one which registers current temperature, and two which have been pushed by this hand to maximum and minimum positions. This type of probe is ideal for soil temperatures, and it costs only $25, but each instrument must be checked for accuracy against a U tube before using; some of them are not accurate.

Temperature can be continuously recorded with instruments called thermographs. Contraction or expansion of metal sensors is translated to movement of an ink pen on chart paper. The paper moves on a drum which is driven by batteries or a spring. Daily, weekly, or monthly clocks are available. Some thermographs use mercury-filled probes on long leads, which may be inserted in water or soil, and others record humidity at the same time. In any case, such instruments must be kept in a shelter. The various max-mim thermometers should also be kept out of direct exposure.

XIX. WEATHER INSTRUMENT SHELTER

The objective of an instrument shelter is to allow free circulation of air between inside and outside, yet prevent exposure of the instruments to direct sunlight or rain. A standard shelter is a wooden box 30 inches wide, 20 inches deep, and 32 inches high. All four sides are louvered, and there is a double roof, with air space between, to prevent heating. All surfaces are white, to reflect heat.

Standard procedure calls for the shelter to be mounted on a four-legged base 48 inches above the ground, and placed in an open area away from obstructions, with the hinged shelter door facing north. The legs must be anchored to prevent overturning in high wind.

Selected
Bibliography

GEOLOGY.

Bird, E. C. F. 1969. *Coasts*. Cambridge, Mass.: M.I.T. Press.

Bullard, E. 1969. The origin of the oceans. In *The Ocean, a Scientific American Book*, pp. 16–25. San Francisco: W. H. Freeman.

Dickinson, W. R., and Grantz, A., eds. 1968. Proccedings of conference on geologic problems of San Andreas Fault System. *Stanford University Publications in Geological Sciences* 11: 1–374.

Higgins, C. G. 1952–4. Lower course of the Russian River, California. *University of California Publications in Geological Sciences* 29: 181–263.

Johnson, F. A. 1948. Petaluma region. *California Department of Natural Resources, Division of Mines Bulletin* 118: 622–627.

Koonig, J. B. 1963. The geologic setting of Bodega Head. *Mineral Information Service, State of California, Division of Mines and Geology* 16(7): 1–9.

BOTANY.

Barbour, M. G. 1970. The flora of plant communities of Bodega Head, California. *Madroño* 20: 289–313.

———. 1972. Additions and corrections to the flora of Bodega Head, California. *Madroño* 21: 446–448.

Billings, W. D. 1964. *Plants and the ecosystem.* Belmont, California: Wadsworth.

Chapman, V. J. 1964. *Coastal vegetation.* New York: Macmillan.

Clausen, J., Keck, D. D.; and Hiesey, W. M. 1940. *Experimental studies on the nature of species.* Carnegie Institute of Washington, Publication No. 520. Washington, D.C.

Goldman, C. R., ed. 1966. *Primary productivity in aquatic environments.* Berkeley: University of California Press.

Hale, M. E., Jr., and Culberson, W. L. 1966. A third checklist of the lichens of the continental United States and Canada. *The Bryologist* 69: 141–182.

Howell, J. T. 1970. *Marin flora.* 2nd ed. Berkeley: University of California Press.

Lewin, R. A., ed. 1962. *Physiology and biochemistry of algae.* New York: Academic Press.

Mason, H. L. 1957. *A flora of the marshes of California.* Berkeley: University of California Press.

Munz, P. A. 1964. *Shore wildflowers of California, Oregon, and Washington.* Berkeley: University of California Press.

Munz, P. A., and Keck, D. D. 1963. *A California flora.* Berkeley: University of California Press.

Orr, R. T., and Orr, D. B. 1968. *Mushrooms and other common fungi of the San Francisco Bay region.* Berkeley: University of California Press.

Salisbury, F. B., and Ross, C. 1969. *Plant physiology.* Belmont, California: Wadsworth.

Smith, G. M. 1969. *Marine algae of the Monterey Peninsula, California.* 2nd ed. Stanford: Stanford University Press.

ZOOLOGY

Berry, W. D., and Berry, E. 1959. *Mammals of the San Francisco Bay Region.* Berkeley: University of California Press.

Borror, D. J., and White, R. E. 1970. *A field guide to the insects of America north of Mexico.* Boston: Houghton Mifflin.

Burt, W. H., and Grossenheider, R. P. 1964. *A field guide to the mammals.* Boston: Houghton Mifflin.

Ingles, L. G., 1965. *Mammals of the Pacific states*. Stanford: Stanford University Press.

Jackson, R. M., and Raw, F. 1966. *Life in the soil*. London: Edward Arnold.

Light, S. F.; Smith, R. I.; Pittelka, F. A.; Abbott, D. P.; and Weesner, F. M. 1967. *Intertidal invertebrates of the central California coast*. Rev. ed. Berkeley: University of California Press.

MacGinitie, G. E., and MacGinitie, N. 1968. *Natural history of marine animals*. New York: McGraw-Hill, Inc.

Martin, A. C.; Zim, H. S.; and Nelson, A. L. 1951. *American wildlife and plants, guide to wildlife food habits*. New York: Dover.

Peterson, R. T. 1961. *A field guide to western birds*. Boston: Houghton Mifflin.

Ricketts, E. F.; Calvin, J.; and Hedgpeth, J. W. 1968. *Between Pacific tides*. 4th ed., rev. Stanford: Stanford University Press.

Smith, A. C. 1959. *Natural history of the San Francisco Bay region*. Berkeley: University of California Press.

Sokal, R. R., and Sneath, P. H. A. 1963. *Principles of numerical taxonomy*. San Francisco: W. H. Freeman.

Stebbins, R. C. 1966. *A field guide to western reptiles and amphibians*. Boston: Houghton Mifflin.

GENERAL ECOLOGY.

Cox, G. W. 1967. *Laboratory manual of general ecology*. Dubuque, Iowa: W. C. Brown.

Detwyler, T. R., ed. 1971. *Man's impact on environment*. San Francisco: McGraw-Hill.

Goodman, G. T.; Edwards, R. W.; and Lambert, J. M., eds. 1965. *Ecology and the industrial society*. New York: John Wiley and Sons.

Hedgpeth, J. W., ed. 1957. Treatise on marine ecology and paleoecology. *Memoirs of the Geological Society of America* 67: 1–1296.

Odum, E. P. 1971. *Fundamentals of ecology*. 3rd ed. Philadelphia: W. B. Saunders.

Phillipson, J. 1966. *Ecological energetics*. London: Edward Arnold.

Smith, R. L. 1966. *Ecology and field biology*. New York: Harper and Row.

Southwood, T. R. E. 1966. *Ecological methods*. London: Methuen.

Whittaker, R. H. 1970. *Communities and ecosystems*. New York: Macmillan.

HISTORY AND ARCHAEOLOGY.

Heizer, R. F., and Whipple, M. A., eds. 1951. *The California Indians*. Berkeley: University of California Press.

Holdren, J., and Herrera, P. 1971. *Energy*. San Francisco: Sierra Club.

Hoover, M. B.; Rensch, H. E.; and Rensch, E. G. 1948. *Historic spots in California*. Stanford: Stanford University Press.

Kinnard, L. 1966. *History of the greater San Francisco Bay region*. 3 vols. New York: Lewis Historical Publ. Co.

Literature Cited

Addicott, W. O. 1952. Ecological and natural history of the pelecypod genus *Macoma* in Elkhorn slough, California. Unpublished M.A. thesis, Stanford University.

Anderson, D. L. 1972. The San Andreas Fault. *Continents Adrift*, W. H. Freeman Co., San Francisco: 142–157.

Barbour, M. G. 1970a. Germination and early growth of the strand plant *Cakile maritima*. *Bulletin of the Torrey Botanical Club* 97: 13–22.

———. 1970b. Is any angiosperm an obligate halophyte? *American Midland Naturalist* 84: 105–120.

———. 1970c. Seedling ecology of *Cakile maritima* along the California coast. *Bulletin of the Torrey Botanical Club* 97: 280–289.

———. 1972. Seedling establishment of *Cakile maritima* at Bodega Head, California. *Bulletin of the Torrey Botanical Club* 99: 11–16.

Barbour, M. G., and Davis, C. B. 1970. Salt tolerance of five California salt marsh plants. *American Midland Naturalist* 84: 262–265.

Barbour M. G., and Racine, C. H. 1967. Construction and performance of a temperature-gradient bar and chamber. *Ecology* 48: 861–863.

Barbour, M. G., and Rodman, J. E. 1970. Saga of the West Coast sea rockets, *Cakile edentula* spp. *californica* and *C. Maritima. Rhodora* 72: 370–386.

Barrett, S. A. 1908. The ethno-geography of the Pomo and neighboring Indians. *American Archeology and Ethnology* 6: 1–332.

Bird, J. B. 1970. Paleo-Indian discoidal stones from southern South America. *American Antiquity* 35: 205–209.

Bitman, J.; Cecil, H. C.; Harris, S. J.; and Fries, G. F. 1969. DDT induces a decrease in eggshell calcium. *Nature* 224: 44–46.

Blus, L. J. 1970. Measurements of brown pelican eggshells from Florida and South Carolina. *BioScience* 20: 867–869.

Boyce, S. G. 1954. The salt spray community. *Ecological Monographs* 24: 29–67.

Brown, V., and Andrews, D. 1969. *The Pomo Indians of California and their neighbors.* Healdsburg, California: Naturegraph Publishers.

Browning, B. M. 1972. The natural resources of Elkhorn Slough: their present and future use. State of California Dept. Fish and Game, Sacramento.

Broyer, T. C., Carlton, A. B.; Johnson, C. M.; and Stout, P. R. 1954. Chlorine, a micronutrient element for higher plants. *Plant Physiology* 29: 526–532.

Calhoun, J. B., and Casby, J. N. 1958. *The calculation of home and density of small mammals.* U.S. Dept. of Health, Education, and Welfare. Public Health Monograph No. 1958. Washington, D.C.

Castenholz, R. W. 1961. The effect of grazing on marine littoral diatom populations. *Ecology* 42: 783–794.

Chapman, V. J. 1960. *Salt marshes and salt deserts of the world.* New York: Interscience Publishers.

Clements, F. E. 1916. *Plant succession: an analysis of the development of vegetation.* Carnegie Institution of Washington, Pub. 242, Washington, D.C.

Coan, E. V. 1971. The northwest American Tellinidae. *Veliger* 14: supplement.

Colley, C. C. 1970. The missionization of the Coast Miwok Indians of California. *The California Historical Society Quarterly* 49: 143–162.

Colman, F. H. 1967. Our walkabout eucalypt. *Walkabout, Australia's way of life magazine* 33 (11): 24–27.

Colman, J. 1933. The nature of the intertidal zonation of plants and animals. *Journal of the Marine Biological Association* 18: 435–476.

Connell, J. H. 1961. The influence of interspecific competition and other factors on the distribution of the barnacle *Chthamalus stellatus. Ecology* 42: 710–723.

———. 1970. A predator-prey system in the marine intertidal region. I. *Balanus glandula* and several predatory species of *Thais. Ecological Monographs* 40: 49–78.

Cook, S. F. 1946-7. A reconsideration of shellmounds with respect to population and nutrition. *American Antiquity* 12: 50–53.

Cox, G. W. 1967. *Laboratory manual of general ecology.* Dubuque, Iowa; W. C. Brown.

Cubit, J. 1969. Behavior and physical factors causing migration and aggregation of the sand crab *Emerita analoga* (Stimpson). *Ecology* 50: 118–123.

Davis, L. V., and Gray, I. E. 1966. Zonal and seasonal distribution of insects in North Carolina salt marshes. *Ecological Monographs* 36: 275–295.

Dayton, P. 1971. Competition, disturbance and community organization: the provision and subsequent utilization of space in a rocky intertidal community. *Ecological Monographs* 41: 351–389.

DeLong, K. T. 1966. Population ecology of feral house mice, interference by *Microtus. Ecology* 47: 481–484.

Dore, W. G. 1958. A simple chemical light meter. *Ecology* 39: 151–152.

Doty, M. S. 1946. Critical tide factors that are correlated with the vertical distribution of marine algae and other organisms along the Pacific coast. *Ecology* 27: 315–328.

Drysdale, F. R. 1971. *Studies in the biology of* Armeria maritima *(Mill.) Willd. with emphasis upon the variety* californica *(Boiss.) as it occurs at Bodega Head, Sonoma County, Cali-*

fornia. Ph.D. dissertation, University of California, Davis.

Eber, L. E.; Saur, J. F. T.; and Sette, O. E. 1968. *Monthly mean charts of sea surface temperature, North Pacific Ocean, 1948–62*. U.S. Dept. of the Interior Circular 258. Washington, D.C.

Ewart, G. Y., and Baner, L. D. 1950. Salinity effects on moisture-electrical resistance relationships. *Soil Science of America Proceedings* 15: 56–63.

Fisher, W. K., and MacGinitie, G. E. 1928. The natural history of an echiuroid worm. *Annals of the Magazine of Natural History*. Series 10, 1: 204–213.

Foster, A. S., and Gifford, E. M. 1959. *Comparative morphology of vascular plants*. San Francisco: W. H. Freeman.

Frank, P. W. 1965. The biodemography of an intertidal snail population. *Ecology* 46: 831–844.

Gates, D. M. 1965. Energy, plants, and ecology. *Ecology* 46: 1–13.

Gause, G. F. 1934. *The struggle for existence*. Baltimore: Williams and Wilkins.

Ghiselin, M. T. 1974. The Economy of Nature and the Evolution of Sex. Berkeley: University of California Press (in press).

Giesel, J. T. 1970. On the maintenance of a shell pattern and behavior polymorphism in *Acmaea digitalis*, a limpet. *Evolution* 24: 98–119.

Gifford, E. W. 1948-9. Diet and the age of Californian shellmounds. *American Antiquity* 14: 223–224.

Giguere, P. E. 1970. The natural resources of Bolinas Lagoon: their status and future. California Dept. Fish and Game, Sacramento.

Glassow, M. A. 1967. Considerations in estimating prehistoric California coastal populations. *American Antiquity* 32: 354–359.

Gleason, H. A. 1926. The individualistic concept of the plant association. *Bulletin of the Torrey Botanical Club* 53: 7–26.

———. 1939. The individualistic concept of the plant association. *American Midland Naturalist* 21: 92–110.

Glynn, P. W. 1965. Community composition, structure and interrelationships in the marine intertidal *Endocladia muricata-Balanus glandula* association in Monterey, California. *Beaufortia* 12: 1–198.

Goldman, C. R. 1960. Molybdenum as a factor limiting primary productivity in Castle Lake, California. *Science* 132: 1016–1017.

Golley, F. B. 1960. Energy dynamics of a food chain in an old field community. *Ecological Monographs* 30: 187–206.

———. 1961. Energy values of ecological materials. *Ecology* 42: 581–584.

Greengo, R. E. 1951. Molluscan species in California shell middens. *Reports of the University of California Archeological Survey*, No. 13.

Hairston, N. G.; Smith, F. E.; and Slobodkin, L. B. 1960. Community structure, population growth and competition. *American Naturalist* 879: 421–425.

Hardin, G. 1960. The competitive exclusion principle. *Science* 131: 1292–1297.

Harper, J. L. 1961. Approaches to the study of plant competition. *Symposium of the Society of Experimental Biology* 15: 1–39.

Harris, G. A. 1967. Some competitive relationships between *Agropyron spicatum* and *Bromus tectorum*. *Ecological Monographs* 37: 89–111.

Hawkes, J., ed. *The world of the past*. New York: Alfred A. Knopf.

Heath, R. G.; Spann, J. W.; and Kreitzer, J. F. 1969. Marked DDE impairment of mallard reproduction in controlled studies. *Nature* 224: 47–48.

Hedgpeth, J. W., ed. 1957. Treatise on marine ecology and paleoecology. *Memoirs of the Geological Society of America* 67: 1–1296.

Heizer, R. F. 1952. A review of problems in the antiquity of man in California. *Reports of the University of California Archeological Survey* No. 16: 3–10.

Heizer, R. F., and M. A. Whipple, eds. 1951. *The California Indians*. Berkeley: University of California Press.

Hewatt, W. G. 1937. Ecological studies on selected marine intertidal communities of Monterey, California. *American Midland Naturalist* 18: 161–206.

Hickey, L. J., and Anderson, D. W. 1968. Chlorinated hydrocarbons and eggshell changes in raptorial and fish-eating birds. *Science* 162: 271–273.

Howell, J. T. 1939. A Russian collection of California plants. Leaflets of *Western Botany* 2: 17–20.

Hunt, R. D., and Sanchez, N. 1929. *A short history of California.* New York: Thomas Y. Crowell.

Jackson, M. L. 1958. *Soil chemical analysis.* Englewood Cliffs, N.J.; Prentice-Hall.

Johansen, H. W. 1966. The benthic marine algae of Bodega Head. In *A marine ecological survey of the Bodega Head region,* pp. 35–112, submitted by Cadet Hand to the University of California, report UCB-34P-96-1.

Johnson, R. G. 1970. Variations in diversity within benthic marine communities. *American Naturalist* 104: 285–300.

———. 1971. Animal-sediment relations in shallow water benthic communities. *Marine Geology* 11: 93–104.

———. 1972. Conceptual models of benthic marine communities. In *Models in paleobiology,* pp. 148–159, T. J. M. Schopf, ed., San Francisco, Freeman Cooper and Company.

Jolly, G. M. 1965. Explicit estimates from capture-recapture data with both death and immigration. *Biometrika* 52: 225–247.

Kershaw, K. A. 1964. *Quantitative and dynamic ecology.* London: Edward Arnold.

King, J. A. 1968. Biology of *Peromyscus. American Society of Mammalogists,* Special Publication No. 2: 375–381.

Knox, W. K. 1915. Nitrogen and chlorine in rain and snow. *Chemistry News* 111: 61–62.

Krebs, C. J. 1966. Demographic changes in fluctuating populations of *Microtus californicus. Ecological Monographs* 36: 239–273.

Kumler, M. L. 1969. Plant succession on the sand dunes of the Oregon coast. *Ecology* 50: 695–704.

Leighton, D. L. 1965. Giant kelp and sea urchins. In *Proceedings of the Fifth International Seaweed Symposium,* pp. 141–153. Pergamon, N.Y.: Pergamon Press.

Lesko, G. L., and Walker, R. B. 1969. Effect of sea water on seed germination in two Pacific atoll beach species. *Ecology* 50: 730–734.

Lewis, J. R. 1964. *The ecology of rocky shores.* London: English Universities Press.

Lidicker, W. Z., Jr. 1966. Ecological observations on a feral house mouse population declining to extinction. *Ecological Monographs* 36: 27–50.

Lindeman, R. L. 1942. The trophic-dynamic aspect of ecology. *Ecology* 23: 399–418.

MacArthur, R. H. 1968. The theory of the niche. In *Population biology and evolution*, edited by R. C. Lewontin, pp. 159–176. Syracuse, New York: Syracuse University Press.

Macdonald, K. B. 1969. Quantitative studies of salt marsh faunas from the North America Pacific coast. *Ecological Monographs* 39: 33–60.

MacGinitie, G. E. 1935. Ecological aspects of a California marine estuary. *American Midland Naturalist* 16: 629–765.

MeGeein, D. J., and Mueller, W. C. 1955–6. A shellmound in Marin County, California. *American Antiquity* 21: 52–62.

McLean, J. H. 1962. Sublittoral ecology of kelp beds of the open coast area near Carmel, California. *Biology Bulletin* 122: 95–114.

McNaughton, S. J. 1966. Thermal inactivation properties of enzymes from *Typha latifolia* L. ecotypes. *Plant Physiology* 41: 1736–1738.

———. 1967. Photosynthetic system. II: Racial differentiation in *Typha latifolia*. *Science* 156: 1363.

Mason, H. L. 1957. *A flora of the marshes of California*. Berkeley: University of California Press.

Millar, C. E.; Turk, L. M. and Foth, H. D. 1958. *Fundamentals of soil science*. 3rd ed. New York: John Wiley and Sons.

Morton, M. L. 1967. The effects of insolation on the diurnal feeding pattern of white-crowned sparrows (*Zonotrichia leucophrys gambelii*). *Ecology* 48: 690–694.

Murray, K. F. 1965. Population changes during the 1957–1958 vole (*Microtus*) outbreak in California. *Ecology* 46: 163–171.

Nelkin, D. 1971. *Nuclear power and its critics*. Ithaca, New York: Cornell University Press.

Neuschull, M. 1967. Studies of subtidal marine vegetation in western Washington. *Ecology* 48: 83–94.

North, W. J. 1963. Ecology of the rocky nearshore environment in southern California. *International Journal of Air and Water Pollution* 7: 721–736.

Oberlander, G. T. 1956. Summer fog precipitation in San Francisco peninsula. *Ecology* 37: 851.

Odum, H. T. 1957. Trophic structure and productivity of Silver Springs, Florida. *Ecological Monographs* 27: 55–112.

Oosting, H. J. 1942. An ecological analysis of the plant communities of Piedmont, North Carolina. *American Midland Naturalist* 28: 1–126.

———. 1956. *The study of plant communities.* 2nd ed. San Francisco: W. H. Freeman.

Oosting, H. J., and Billings, W. D. 1942. Factors affecting vegetation zonation on coastal dunes. *Ecology* 23: 131–142.

Osborne, D. 1958. Western American prehistory—an hypothesis. *American Antiquity* 24: 47–52.

Paine, R. T. 1969. The *Pisaster-Tegula* interaction. *Ecology* 50: 950–961.

Phillips, E. A. 1959. *Methods of vegetation study.* New York: Holt, Rinehart, and Winston.

Porter, R. D., and Wiemeyer, S. N. 1969. Dieldrin and DDT, effects on sparrow hawk eggshells and reproduction. *Science* 165: 199–200.

Ratcliffe, D. A. 1967. Decrease in eggshell weight in certain birds of prey. *Nature* 215: 208–210.

Recher, H. F. 1966. Some aspects of the ecology of migrant shore birds. *Ecology* 47: 393–407.

Rhoades, D. C., and D. K. Young. 1970. The influence of deposit-feeding organisms on sediment stability and community structure. *Journal of Marine Research* 28: 150–178.

Richards, L. A., ed. 1954. *Diagnosis and improvement of saline and alkali soils.* U.S. Dept. of Agriculture Handbook No. 60. U.S. Government Printing Office, Washington, D.C.

Risebrough, R. W.; Huggett, R. J.; Griffin, J. J.; and Goldberg, E. D. 1968. Pesticides: transatlantic movements in the northeast trades. *Science* 159: 1233–1236.

Robbins, W. W. 1940. *Alien plants growing without cultivation in California.* University of California Agricultural Experiment Station Bulletin No. 637: 1–127.

Robbins, W. W.; Bellue, M. K.; and Ball, W. S. 1951. *Weeds of California.* Sacramento: State of California Printing Division.

Rudd, R. L. 1966. *Pesticides and the living landscape.* Madison: University of Wisconsin Press.

———. 1970. Chemicals in the environment. *California Medicine* 113: 27–32.

Salt, G. W., and Willard, D. E. 1971. The hunting behavior and success of Forster's terns. *Ecology* 52: 989–998.

Schoener, T. W. 1968. Sizes of feeding territories among birds. *Ecology* 49: 123–141.

Schwarz, M. 1970. "A lost generation of pelicans, man's fault." *San Francisco Chronicle*, 26 August 1970, p. 3.

Shepard, F. P. 1964. Sea level changes in the past 6000 years: possible archeological significance. *Science* 143: 574–576.

Sladen, W. J. L.; Menzie, C. M.; and Reichel, W. L. 1966. DDT residues in Adelie penguins and a crabeater seal from Antarctica. *Nature* 210: 670–673.

Smith, G. M. 1945. The marine algae of California. *Science* 101: 188–192.

Snedecor, G. W., and Cochran, W. G. 1967. *Statistical methods.* 6th ed. Ames, Iowa: Iowa State University Press.

Soil Survey Staff. 1951. *Soil survey manual.* U.S. Dept. of Agriculture Handbook No. 18. U.S. Government Printing Office, Washington, D.C.

Sokal, R. R., and Rohlf, F. J. 1969. *Biometry.* San Francisco: W. H. Freeman.

Southwood, T. R. E. 1966. *Ecological methods.* London: Methuen.

Spoecker, P. 1966. An analysis of settling surfaces at two sites on Bodega Head. In *A marine ecological survey of the Bodega Head region,* pp. 1–15, submitted by Cadet Hand to the University of California, report UCB-34P-96-1.

Stephenson, T. A., and Stephenson, A. 1949. The universal features of zonation between tides marks on rocky coasts. *Journal of Ecology* 37: 289–305.

Stone, E. C., and Fowells, H. A. 1955. Survival value of dew under laboratory conditions with *Pinus ponderosa. Forest Science* 1: 183–188.

Sutherland, J. P. 1970. Dynamics of high and low populations of the limpet, *Acmaea scabra* (Gould). *Ecological Monographs* 40: 169–188.

Test, A. R. 1946. Speciation in limpets of the genus *Acmaea. Contributions from the Laboratory of Vertebrate Biology* 31: 1–24.

Tilly, L. J. 1968. The structure and dynamics of Cone Spring. *Ecological Monographs* 38: 169–197.

Turesson, G. 1922. The genotypical response of the plant species to the habitat. *Hereditas* 3: 211–350.

Turill, W. B. 1946. The ecotype concept. *The New Phytologist* 45: 34–43.

Ulrich, A., and Ohki, K. 1956. Chlorine, bromine, and sodium as nutrients for sugar beet plants. *Plant Physiology* 31: 171–181.

U.S. Department of Commerce. 1968. *Climatological data, United States by sections.* Vol. 55. Washington, D.C.

U.S. Department of Interior. 1970. FWPCA issues report on environmental impacts of coastal power plants. *Clean Water News,* April 16. Washington, D.C.

Vogelmann, H. W.; Siccama, T.; Leedy, D.; and Ovitt, D.C. 1968. Precipitation from fog moisture in the Green Mountains of Vermont. *Ecology* 49: 1205–1207.

Walker, K. R. 1972. Trophic analysis: a method for the studying of ancient communities. *Journal of Paleontology* 46: 82–93.

Warme, J. E. 1971. Paleoecological aspects of a modern coastal lagoon. *University of California Publications in Geology* 87: 133 p.

Went, F. W. 1970. Climate and agriculture. In *Plant agriculture, a Scientific American Book,* pp. 108–118. San Francisco: W. H. Freeman.

Williams, W. A. 1966. Range improvement as related to net productivity, energy flow, and foliage configuration. *Journal of Range Management* 19: 29–34.

Williams, W. T., and Potter, J. R. 1972. The coastal strand community at Morro Bay State Park, California. *Bulletin of the Torrey Botanical Club* 99: 163–171.

Williams, W. T. 1973. Species dynamism in the Pacific coastal strand community at Morro Bay, California. *Bulletin of the Torrey Botanical Club* (in press).

Wolcott, T. 1969. *Physiological ecology of limpets (Acmaeidae) in the rocky intertidal at Bodega Bay, California.* Ph.D. dissertation, University of California, Berkeley.

Wurster, D. H., C. F. Wurster, and W. N. Strickland. 1965. Bird mortality following DDT spray for dutch elm disease. *Ecology* 46: 488–499.

Zeller, R. W., and Rulifson, R. L. 1970. *A survey of California coastal power plants.* Federal Water Pollution Control Administration, Northwest Region. Portland, Oregon.

Author Index

This index only includes those authors cited by name in the text; it does not include all the entries listed in "Selected Bibliography" or "Literature Cited."

Subject Index

Abalone, red. See *Haliotis rufescens*
Abronia latifolia (sand verbena), 144, 146, 148, 265
Acaena californica, 267
Acanthina (thorn snail), 116
Achillea borealis (yarrow): ground cover in dunes, 144; in disturbed areas, 223, 258
Acmaea asmi (limpet), 117
Acmaea digitalis (limpet): habitat, 114, 117; taxonomy, 119
Acmaea insessa (limpet), 118
Acmaea instabilis (limpet), 122
Acmaea limatula (limpet), 19
Acmaea mitra (limpet), 118
Acmaea paleacea (limpet), 122
Acmaea paradigitalis (limpet), 117
Acmaea pelta (limpet), 99, 117, 120
Acmaea persona (limpet), 114
Acmaea scabra (limpet): habitat, 114, 117; population dynamics, 114
Acmaea scutum (limpet), 117
Actitis macularia (spotted sandpiper), 102, 278

Aechmorphorus occidentales (western grebe): habitat, 187, 275; effect of pesticides on, 224–225
Aestivation, 73
Agelaius phoenicius (red-winged blackbird), 81, 281
Aglaja, 181
Agoseris apargioides (beach dandelion), 146, 258
Agrostis alba (red top), 271
Agrostis exarata (bent grass), 271
Aira caryophyllea (hair grass), 59, 271
Alchemilla occidentalis, 267
Alderia (sea slug), 173
Algae: diversity, 18; in salt marsh, 173; in mud flats, 158, 159, 162, 175, 180–182; in fresh-water marsh, 204; in rocky intertidal, 113, 117, 118, 122; as Indian food, 242. See also *Egregia; Enteromorpha; Laminaria; Oedogonium; Pelvetiopsis; Postelsia; Spirogyra; Ulva*
Alligator lizard, northern. See *Gerrhonotus coerleus*

155 ff; mud flats, 173 ff; fresh-water marsh, 196 ff; disturbed areas, 213 ff.
Habitat ecology, defined, 1; methods of, 24 ff
Haematopus bachmani (black oyster catcher), 102, 277
Hair grass. See *Aira caryophylla*
Hairy cat's ear. See *Hypochoeris radicata*
Haliotis rufescens (red abalone): habitat, 100, 121; and sea otter, 127; as Indian food, 243
Haminoea (bubble shell snail), 181
Haplopappus ericoides: habitat, 260; ground cover in dunes, 144, 146
Harmothoe adventor (scale worm), 179
Hawk, marsh. See *Circus cyaneus*
Hawk, red-tailed. See *Buteo jamaicensis*
Hawk, sparrow. See *Falco sparverius*
Hedge nettle. See *Stachys rigida*
Hedgpeth, J., participant in power plant controversy, 231, 233
Heliotrope, wild. See *Phacelia distans*
Helminthoglypta arrosa (lined snail): seasonality, 73, 74; in food web, 79, 82, 85; in fresh-water marsh, 207; in middens, 244; as Indian food, 242
Hemerocampa vetusta (California tussock moth), 72
Hemigrapsus nudus (purple shore crab): in rocky intertidal, 121; in salt marsh, 173
Hemigrapsus oregonensis (shore crab), 173
Hemlock, poison. See *Conium maculatum*
Hepialus behrensi, 73
Heracleum lanatum (cow parsnip): part of northern coastal scrub, 67, 270; on shaded banks, 211; as Indian food, 242
Herbivore, defined, 71
Heron, black-crowned night. See *Nycticorax nycticorax*
Heron, great blue. See *Ardea herodias*
Hermissenda crassicornis, 181

Heteroscelus incanus (wandering tattler), 102, 278
Hibernation, defined, 73
Himalaya berry. See *Rubus procerus*
Hirundo rustica (barn swallow), 190, 280
Hitchcock, A., directed filming of film *The Birds*, 205
Holcus lanatus (velvet grass): in fresh-water marsh, 201, 202; in depressions, 207; in salt marsh, 207; description, 202, 272
Hordeum brachyantherum, 272
Hordeum depressum (wild barley), 272
Hordeum leporinum (farmer's foxtail), 59, 272
Horehound. See *Marrubium vulgare*
Horkelia marinensis, 268
Horsetail. See *Equisetum*
Horseweed. See *Conyza canadensis*
Hottentot fig. See *Mesembryanthemum edule*
House finch. See *Carpodacus mexicanus*
Hukueko. See Indian
Hummingbird, Anna's. See *Calypte anna*
Hydrocotyle ranunculoides (marsh pennywort), 204, 270
Hydrophroghia caspia (Caspian tern), 188, 279
Hyla regilla (Pacific tree frog): habitat, 82, 274; in food web, 81, 82
Hypochoeris radicata (hairy cat's ear): habitat, 260; in fresh-water marsh, 201–202; in disturbed areas, 223

Ice plant, sea fig. See *Mesembryanthemum chilense*
Idothea, 180
Indians: age of, 213; artifacts, 6, 241–242; native Indian groups of Bodega Head, 240–241; Olamentko and Hukueko dialects, 239–41; population size, 241, 247; villages at Bodega Head, 240–241; food sources, 242–245; protein intake, 243; missionization of, 247; work/food ratio, 245–246